National Institute of
Standards and Technology
Technology Administration
U.S. Department of Commerce

NIST Special Publication 500-262

Proceedings of the Static Analysis Summit

Paul E. Black, Helen Gill, and W. Bradley Martin (co-chairs)
Elizabeth Fong (editor)

Information Technology Laboratory
National Institute of Standards & Technology
Gaithersburg MD 20899

July 2006

U.S. Department of Commerce
Carlos M. Gutierrez. Secretary

National Institute of Standards and Technology
William Jeffrey, Director

Disclaimer: Any commercial product mentioned is for information only; it does not imply recommendation or endorsement by NIST nor does it imply that the products mentioned are necessarily the best available for the purpose.

Proceedings of the Static Analysis Summit

Paul E. Black, Helen Gill, and W. Bradley Martin (co-chairs)
Elizabeth Fong (editor)

Information Technology Laboratory
National Institute of Standards and Technology
Gaithersburg, MD 20899

ABSTRACT

These are the proceedings of a summit held in June 2006 at the National Institute of Standards and Technology (NIST). This Static Analysis Summit is one of a series of meetings in the NIST Software Assurance Measurement and Tool Evaluation (SAMATE) project. This summit convened researchers, developers, and government and industrial users to explore the state of the art in software static analysis tools and techniques with an emphasis on software security. It is also served as a prelude to an international summit in Spring 2007. This proceeding includes the ten papers presented, the keynote presentation, and discussion of a next summit.

Keywords: Software assessment tools; software assurance; software metrics; software security; source code analysis, static analysis, vulnerability.

Foreword

These are the proceedings of the Static Analysis Summit held June 29, 2006, at the National Institute of Standards and Technology (NIST), Gaithersburg, Maryland, USA. The summit was organized in part by the Software Diagnostics and Conformance Testing Division, in NIST's Information Technology Laboratory. These proceedings have four main parts:
- Call for Papers,
- Summit Agenda,
- Presentation Accompanying Keynote Address, and
- Papers

This summit is one of a series of meetings in conjunction with the NIST Software Assurance Measurement and Tool Evaluation (SAMATE) project http://samate.nist.gov/ The SAMATE project is partially funded by DHS to help identify and enhance software security assurance tools. Two previous workshops were conducted: first, "Defining the State of the Art in Software Security Tools," held in August 2005 at NIST in Gaithersburg, Maryland, and second, "Software Security Assurance Tools, Techniques, and Metrics," held in November 2005 in Long Beach, California, USA.

The goal of this summit is to convene researchers, developers, and government and industrial users to explore the state of the art in software static analysis tools and techniques with an emphasis on software security. It is also to serve as a prelude to an international summit in Spring 2007.

The call for papers resulted in ten accepted papers, which were presented at the summit. Professor Dawson Engler, Stanford, gave a keynote address. Sixty people attended from government, universities, tool vendors and service providers, research companies, and industry. Attendees from outside the USA came from the UK and Canada.

The final session discussed future summit meetings. The sentiment was that this summit is too short. The next one should be at least two days long, possibly with breakout groups by language or level for more focused discussions. The next summit should have more people from outside the USA. It should also include mission critical groups, the safety community, and more academicians.

We are especially grateful to Prof. Dawson Engler for his enlightening keynote address. I thank those who worked to organize this summit, particularly my two co-chairs: Helen Gill, NSF and W. Bradley Martin, NSA. We appreciate the program committee for their efforts in reviewing the papers. Many thanks are due to NIST, especially the Software Diagnostics and Conformance Testing division, for providing the organizers' time. On behalf of the program committee and the whole SAMATE team, thanks to everyone for taking their time and resources to join us.

Dr. Paul E. Black
18 July 2006

Table of Contents

Call For Papers 6

Summit Agenda 8

Keynote Presentation
    Dawson Engler 9

Secure Coding Standards
    Robert C. Seacord 14

Language Design for Verification
    Rod Chapman and Peter Amey 17

Automated Calculation of Software Behavior with Function Extraction (FX) for Trustworthy and Predictable Execution
    Richard C. Linger, Stacy J. Prowell, and Mark Pleszkoch 22

Support for Whole-Program Analysis and the Verification of the One-Definition Rule in C++
    Dan Quinlan, Richard Vuduc, Thomas Panas, Jochen Hardtlein, and Andreas Saebjornsen 27

Towards the Industrial Scale Development of Custom Static Analyzers
    John Anton, Eric Bush, Allen Goldberg, Klaus Havelund, Doug Smith and Arnaud Venet 36

Verification Tools for Software Security Bugs
    Frederic Michaud and Frederic Painchaud 41

A Framework for Creating Custom Rules for Static Analysis Tools
    Eric Dalci and John Steven 49

High Fidelity Static Analysis for Secure Enterprise Software Requires Platform Knowledge
    Nikolai Mansourov, Djenana Campara, Norman Rajala, and Sumeet Malhotra 55

A Status Update: The Common Weakness Enumeration
    Robert A. Martin and Sean Barnum 62

A Proposed Functional Specification for Source Code Analysis Tools
    Mickael Kass, Michael Koo, Paul E. Black, and Vadim Okun 65

CALL FOR PAPERS

National Institute of Standards and Technology (NIST)
Software Assurance Metrics and Tool Evaluation (SAMATE) Project

Static Analysis Summit

29 June 2006
http://samate.nist.gov/SAS
Gaithersburg, MD, USA

"Black-box" software testing cannot realistically find maliciously implanted Trojan horses or subtle errors which have many preconditions. For maximum reliability and assurance, static analysis must be applied to all levels of software artifacts, from models to source code to byte code to binaries. The goal of this workshop is to convene researchers, developers, and government and industrial users to explore the state of the art in software static analysis tools and techniques with an emphasis on software security.

We solicit contributions describing basic research, novel applications, experience, and proposals relevant to static analysis tools, techniques, and their evaluation. Topics of particular interest are:

- What is possible with today's techniques?
- What is feasible with today's tools?
- What is NOT possible or feasible with current tools or techniques?
- Where are the gaps that further research might fill?
- What is the minimum performance bar for a source code analyzer?
- Static analysis' contribution to software security assurance
- Flaw catching effectiveness of methods, techniques, or tools
- Benchmarks or reference datasets
- Software security assurance metrics
- How can users, developers, or researchers evaluate the performance of static analysis tools?
- User experience drawing useful lessons or comparisons.

SUBMISSIONS:

Papers should be from 1 to 8 pages long. Papers exceeding eight pages will not be reviewed. All submissions should clearly identify their novel contributions.

Submit papers electronically in PDF or ASCII text by 20 May 2006 to Liz Fong <efong@nist.gov>. Your submission constitutes permission for us to publish it in workshop proceedings.

We will notify submitters of acceptance by 1 June 2006.

PUBLICATION:

Accepted papers, along with workshop presentations where possible, will be published in the workshop proceedings as a NIST Special Publication.

IMPORTANT DATES:

20 May: Paper submission deadline
1 June: Author notification
13 June: Final camera-ready copy due
29 June: Workshop

ORANIZERS:
Co-Chairs: Paul Black   NIST, paul.black@nist.gov
            Helen Gill   NSF, hgill@nsf.gov
            W. Bradley Martin NSA, wbmarti@tycho.nsa.gov

PROGRAM COMMITTEE:

Freeland Abbott	Georgia Tech	Paul Ammann	George Mason U.
Paul Anderson	GrammaTech	John Anton	Kestrel
Ira Baxter	Semantic Designs	Rogier Boon	Itsec Security
Djenana Campara	KDM Analytics	Pravir Chandra	Secure Software
Ben Chelf	Coverity	Brain Chess	Fortify
Jack Danahy	Ounce Labs	Elizabeth Fong	NIST
Larry Johnsen	Parasoft	Michael Kass	NIST
Michael Koo	NIST	Robert E. Lee	GMRI
Robert A. Martin	MITRE Corp.	Vadim Okun	NIST
Daniel J. Quinlan	LLNL	Ioana Rus	Fraunhofer USA
Ravi Sandhu	George Mason U.	Robert C. Seacord	CERT/CC

LOCAL ARRANGEMENTS:

Elizabeth Fong   NIST, efong@nist.gov

Summit Agenda

8:30 - 9:00 : Registration

9:00 - 9:30 : Welcome - Cita Furlani, Director, Information Technology Laboratory, NIST
    * Program Presentation and Charge to Attendees - Paul E. Black

9:30 - 10:20 :
    * Secure Coding Standards - Robert C. Seacord
    * Language Design for Verification - Rod Chapman and Peter Amey

10:20 - 10:45 : Break

10:45 - 12:00:
* Automated Calculation of Software Behavior with Function Extraction (FX) for Trustworthy and Predictable Execution - Richard C. Linger, Stacy J. Prowell, and Mark Pleszkoch
* Support for Whole-Program Analysis and the Verification of the One-Definition Rule in C++ - Dan Quinlan, Richard Vuduc, Thomas Panas, Jochen Härdtlein, and Andreas Sæbjørnsen
* Towards the Industrial Scale Development of Custom Static Analyzers - John Anton, Eric Bush, Allen Goldberg, Klaus Havelund, Doug Smith, and Arnaud Venet

12:00 - 1:00 : Lunch

1:00 - 1:30 : Keynote: Dawson Engler

1:30 - 2:45 :

    * Verification Tools for Software Security Bugs - Frédéric Michaud and Frédéric Painchaud
* A Framework for Creating Custom Rules for Static Analysis Tools - Eric Dalci and John Steven
* High Fidelity Static Analysis for Secure Enterprise Software Requires Platform Knowledge - Nikolai Mansourov, Djenana Campara, Norman Rajala, and Sumeet Malhotra

2:45 - 3:10 : Break

3:10 - 4:00 :

* A Status Update: The Common Weakness Enumeration - Robert A. Martin and Sean Barnum
  * A Source Code Analysis Tool Specification - Michael Kass and Michael Koo

4:00 - 4:30 :

*The next, international meeting: Where? When? Who else should be invited?

Keynote Presentation – Dawson Engler

Weird things that surprise academics trying to commercialize a static checking tool.

Andy Chou, Ben Chelf, Seth Hallem
Bryan Fulton, Charles Henri-Gros, Scott McPeak, Ted Unangst
Chris Zak
Coverity

Dawson Engler
Stanford

One-slide bio.

- Academic Lineage
  - MIT: PhD thesis = new operating system (exokernel)
  - Stanford: last 7 years developing techniques to find as many serious bugs as possible in large software systems.
  - Co-founded Coverity: 100 customers, cashpositive from T=0
- Our research focuses on three approaches:
  - Implementation-level model checking [OSDI'02, OSDI'04].
  - Automatically generate test cases using symbolic execution [Spin'05, Oakland security'06]
  - System-specific static analysis: use extended compiler to check code. By far the easiest to use and most generally reliable way to find many errors. Rest of the talk on this.

Background: Customized static analysis

- Systems have many ad hoc correctness rules
  - "acquire lock l before modifying x", "cli() must be paired with sti()," "don't block with interrupts disabled"
  - One error = crashed machine
- If we know rules, can check with extended compiler
  - Rules map to simple source constructs
  - Use compiler extensions to express them

```
                save(flags);
Linux           cli();                    EDG compiler
drivers/        if(!(buf = kmalloc()))    ┌──────────────┐    "did not re-
raid5.c             return 0;             │  int checker │     enable ints!"
                restore(flags);           └──────────────┘
                return buf;
```

Nice: scales, precise, statically find 1000s of errors

High bit: Works well.

- A bunch of checkers:
  - System-specific static checking [OSDI'00] (Best paper)
  - Security checkers [Oakland'02] & annotations [CCS'03]
  - Race conditions and deadlocks [SOSP'03]
  - Path-sensitive memory overflows [FSE'03]
  - Others [ASPLOS'00, PLDI'02, PASTE'02, FSE'02(award)]
  - Infer correctness rules [SOSP'01]
  - Z-ranking [SAS'03]
  - Correlation ranking [FSE'03]
- Big system? Always find bugs.
  - New checker, no bugs? Immediate: what's wrong??
- Tenure
- Commercialized(ing): Coverity
  - Successful enough to have a marketing dept.

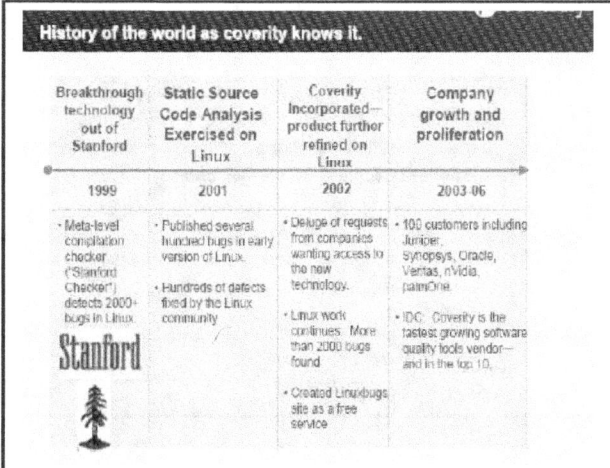

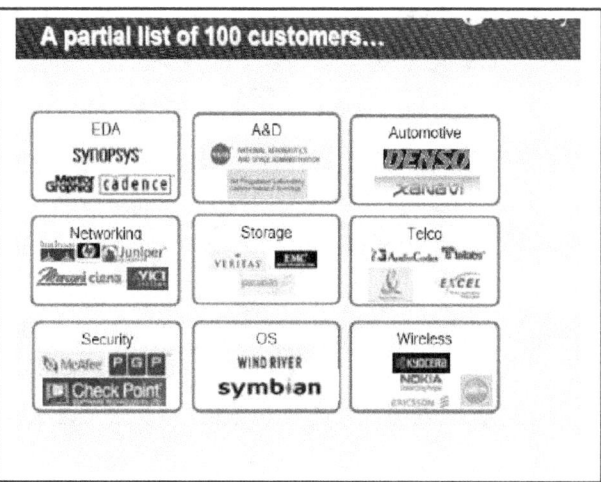

A naïve view

- Initial market analysis:
    "We handle Linux, BSD, we just need a pretty box!"
    Not quite.

- First rule of static analysis: no check, no bug.
    Two first order examples we never would have guessed.
    Problem 1: if you can't find the code, can't check it.
    Problem 2: if you can't compile code, you can't check it.

- And then: how to make money on software tool?
    "Tools. Huh. Tools are hard." Any VC in early 2000.

Myth: the C (or C++) language exists.

- Well, not really. The standard is not a compiler.
    What exists: gcc-2.1.9-ac7-prepatch-alpha, xcc-i-did-not-understand-pages4,33,208-242-of-standard.
    Microsoft: Conformance = competitive disadvantage.
    Basic LALR law: What can be parsed will be written.

- Rule: static analysis must compile code to check.
    If you cannot (correctly) parse "language" cannot check.

    Common (mis)usage model: "allegedly C" header file does something bizarre not-C thing. Included by all source. Customer watches your compiler emit voluminous parse errors.
    Of course: gets way worse with C++ (which we support)

Some bad examples to find in headers

- Banal. But take more time than you can believe:

    `void x;`   `short x; int *y = &(int)x;`   `int foo(int a, int a);`

    `unsigned x @ "TEXT";`   `unsigned x = 0xdead_beef;`

    `Int16 ErrSetJump(ErrJumpBuf buf) = { 0x4E40 + 15, 0xA085 };`

- And, of course, asm:

    ```
    #pragma asm
      mov eax, eab
    #pragma end_asm
    ```

    ```
    asm foo() {
      mov eax, eab;
    }
    ```

    ```
    // newline = end
    __asm mov eax, eab
    ```

    ```
    // "]" = end
    __asm [
       mov eax, eab
    ]
    ```

Microsoft example: precompiled headers

- Spec:

    The compiler treats all code occurring before the .h file as precompiled. It skips to just beyond the #include directive associated with the .h file, uses the code contained in the .pch file, and then compiles all code after filename

- Implication

    I can put whatever I want here.
    It doesn't have to compile.
    If your compiler gives an error it sucks.
    #include <some-precompiled-header.h>

- It gets worse: on-the-fly header fabrication

Solution: pre-preprocessing rewrite rules.

- Supply regular expressions to rewrite bad constructs

    ```
    #pragma asm
    ...
    #pragma end_asm
    ```

    ↓

    ```
    ppp_translate("/#pragma asm/#if 0/");
    ppp_translate("/#pragma end_asm/#endif/");
    ```

    ↓

    ```
    #if 0
    ...
    #endif
    ```

What this all means concretely.

- We use Edison Design Group (EDG) frontend
    Pretty much everyone uses. Been around since 1989.
    Aggressive support for gcc, microsoft, etc. (bug compat!)

- Still: coverity by far the largest source of EDG bugs:
    146 parsing test cases (i.e., we got burned)
    219 compiler line translation test cases (i.e., ibid).
    163 places where frontend hacked ("#ifdef COVERITY")
    Still need custom rewriter for many supported compilers:

    ```
    205 hpux_compilers.c        453 sun_compilers.c
    215 lcr_compiler.c          485 arm_compilers.c
    240 ti_compiler.c           617 gnu_compilers.c
    251 green_hills_compiler.c  748 microsoft_compilers.c
    377 intel_compilers.c      1587 metrowerks_compilers.c
    453 diab_compilers.c        ...
    ```

Annoying amplifier: Can we get source?

- *NO*!
    Despite NDAs
    Even for parse errors
    Even for preprocessed
    Maybe if obfuscated (they probably won't trust).

- Might just be because coverity too small to sue...

- Sales engineer has to type in from memory.
    Of course: usually doesn't replicate problem in isolation...

Academics don't understand money.

- "We'll just charge per seat like everyone else"
    Finish the story: "Company X buys three Purify seats, one for Asian, one for Europe and one for the US..."
- Try #2: "we'll charge per lines of code"
    "That is a really stupid idea: (1) ..., (2) ... , ... (n) ..."
    Actually works. I'm still in shock. Would recommend it.
- Good feature for seller:
    No seat games. Revenue grows with code size. Run on another code base = new sale.
- Good feature for buyer: No seat-model problems
    Buy once for project, then done. No per-seat or per-usage cost; no node lock problems; no problems adding, removing or renaming developers (or machines)
    People actually seem to like this pitch.

(Often) People don't understand much.

- Our initial naïve expectation: People who write code for money understand it. Instead:
    "To build, I just press this button..."
    "I'm just the security guy"
    "That bug is in 3rd party code"
    "Is it a leak? Author left years ago..."
- People also don't understand compilers.
    "Static" analysis? "What is the performance overhead?"
    Anything that finds bugs = testing.
    "Think of it as super compiler warnings"
- They certainly do not understand your tool.
    User not same as tool builder. Uninformed. Inattentive. Cruel.
    Makes difficult to deploy anything sophisticated.
    Example: statistical inference, race conditions.
    In some ways, checkers lag much behind our research ones.

Some commercial experiences

- Surprise: Sales guys are great
    Easy to evaluate. Modular.

- Company X buys "quality improving" tool, then fires 110 people.
    Good or bad?

- Large companies "want" to be honest
    Veritas: want monitoring so don't accidently violate!

- Competitors can really slow down sales cycle.
    Time to sale ~ max(time for all competitors to do a trial).

"No, your tool is broken: that's not a bug"

- "No, the loop will go through once!"
    ```
    for(i=1; i < 0; i++) {
        ...deadcode...
    }
    ```
- "No, && is 'or'!"
    ```
    void *foo(void *p, void *q) {
        if(!p && !q)
            return 0;
    ```
- "No, ANSI lets you write 1 past end of the array!"
    ("We'll have to agree to disagree." !!!)
    ```
    unsigned p[4];  p[4] = 1;
    ```

```
for(s=0; s < n; s++) {
    ...
    switch(s) {
        case 0: assert(0);
            return;
    ...
    }
    ...dead code...
```

Some cursory static analysis experiences

- Getting 1000s of bugs with static *so* *easy* compared to everything else.
    Push button. Compile what you can. Don't have to understand. Bugs (relatively) easy to diagnose.
- Finding errors often easy, saying why is hard
    Have to track and articulate all reasons.
- Ease-of-inspection *crucial*
    Extreme: Don't report errors that are too hard.
- The advantage of checking human-level operations
    Easy for people? Easy for analysis. Hard for analysis? Hard for people.
- Soundness not needed for good results.

Do false positives matter?

- \> 30% false positives kills sale (managers)
    - Users actually accept 70% (or more: security guys).
    - Caveat: First ~3 reports are FPs = "tool sucks"
    - Caveat: Low trust = complex bugs called false positives.
- More important: no embarassing FPs.
    - Stupid FP? Implies tool stupid. Not good for credibility.
    - Social: don't want to embarrass tool champions internally
- More important: no failed merges.
    - Mark FP once? Fine.
    - Reappears and have to mark again? email support.
- More important: inspection time.
    - Even true bugs with high inspection costs not happy.

Do false negatives matter?

- Of course not! Invisible.
- Oops, not always.
    - Upgrade product, set of defects shifts slightly = "Dude, where is my bug?"
    - Or bug gets found some other way.
- Very general problem.
    - Tool has bugs. Some lead to FPs some to FNs.
    - FPs easy to find. Fix.
    - But each fix increases probability that introduce FNs.
    - And doesn't remove FN.
    - Over time: bad deformation.
- Tough call: usually reduce Pr(FN) increases Pr(FP).

Myth: more analysis is always better

- Does not always improve results, and can make worse
- The best error:
    - Easy to diagnose
    - True error
- More analysis used, the worse it is for both
    - More analysis = the harder error is to reason about, since user has to manually emulate each analysis step.
    - Number of steps increase, so does the chance that one went wrong. No analysis = no mistake.
- In practice:
    - Demote errors based on how much analysis required
    - Revert to weaker analysis to cherry pick easy bugs
    - Give up on error classes that are too hard to diagnose.

No bug is too stupid to check for.

- Someone, somewhere will do anything you can think of.
- Best recent example:
    - From security patch for bug found by Coverity in X windows that lets almost any local user get root.

```
--- hw/xfree86/common/xf86Init.c.orig 2006-03-17...
   /* First the options that are only allowed for root */
-  if (getuid() != 0 && geteuid == 0) {
+  if (getuid() != 0 && geteuid() == 0) {
       ErrorF("-configure can only be used by root.\n");
       exit(1);
   }
```

- Next: Two amazingly effective checks.

One of the best stupid checks: Deadcode

- Programmer generally intends to do useful work.
    - Use constraint analysis to flag code where all paths to it are impossible. Often serious logic bug.
- From UU aodv (good code):
    - Linked list removal mistake. After send, take packet off queue. Bug = if any packets on list before the one we want will lose them!

```
// packet_queue.c:packet_queue_send
prev = null;
while(curr) {
    if(curr->dst_addr == dst_addr) {
        if(prev == NULL)
            PQ.head = curr->next;
        else
            ..DEADCODE [prev never updated]..
```

Internal null: trivial, amazingly effective.

- "*p" implies programmer believes p is not null
- A check (p == NULL) implies two beliefs:
    - POST: p is null on true path, not null on false path
    - PRE: p was unknown before check
- Cross-check beliefs: contradiction = error.
- Check-then-use (79 errors, 26 false pos)

```
/* 2.4.1: drivers/isdn/svmb1/capidrv.c */
if(!card)
    printk(KERN_ERR, "capidrv-%d: ..", card->contrnr..)
```

Can look for redundancy in general: deadcode elim is an error finder. Can look for: writes never read, lock acquired that protects nothing. This is more interesting. Contradiction is a symmetric relation.

Null pointer fun

- Use-then-check: 102 bugs, 4 false

```
/* 2.4.7: drivers/char/mxser.c */
struct mxser_struct *info = tty->driver_data;
unsigned flags;
if(!tty || !info->xmit_buf)
    return 0;
```

- Contradiction/redundant checks (24 bugs, 10 false)

```
/* 2.4.7/drivers/video/tdfxfb.c */
fb_info.regbase_virt = ioremap_nocache(...);
if(!fb_info.regbase_virt)
    return -ENXIO;
fb_info.bufbase_virt = ioremap_nocache(...);
/* REDUNDANT check */
if(!fb_info.regbase_virt) {
    iounmap(fb_info.regbase_virt);
```

Assertion: Soundness is often a distraction

- Soundness: Find all bugs of type X.
  Not a bad thing. More bugs good.
  BUT: can only do if you check weak properties.
- What soundness really wants to be when it grows up:
  Total correctness: Find all bugs.
  Most direct approximation: find as many bugs as possible.
- Opportunity cost:
  Diminishing returns: Initial analysis finds most bugs
  Spend time on what gets the next biggest set of bugs
  Easy experiment: bug counts for sound vs unsound tools.
- Soundness violates end-to-end argument:
  "It generally does not make much sense to reduce the residual error rate of one system component (property) much below that of the others."

Static vs dynamic bug finding

- Static: precondition = compile (some) code.
  All paths + don't need to run + easy diagnosis.
  Low incremental cost per line of code
  Can get results in an afternoon.
  10-100x more bugs.
- Dynamic: precondition = compile all code + run
  What does code do? How to build? How to run?
  Runs code, so can check implications.
  Good: Static detects ways to cause error, dynamic can check for the error itself.
- Result:
  Static better at checking properties visible in source, dynamic better at properties implied by source.

Open Q: how to get the bugs that matter?

- Myth: all bugs matter and all will be fixed
  FALSE
  Find 10 bugs, all get fixed. Find 10,000...
- Reality
  All sites have many open bugs (observed by us & PREfix)
  Myth lives because state-of-art is so bad at bug finding
  What users really want: The 5-10 that "really matter"
- General belief: bugs follow 90/10 distribution
  Out of 1000, 100 (10? or 1?) account for most pain.
  Fixing 900+ waste of resources & may make things worse
- How to find worst? No one has a good answer to this.
  Possibilities: promote bugs on executed paths or in code people care about.

Open Q: Do static tools really help?

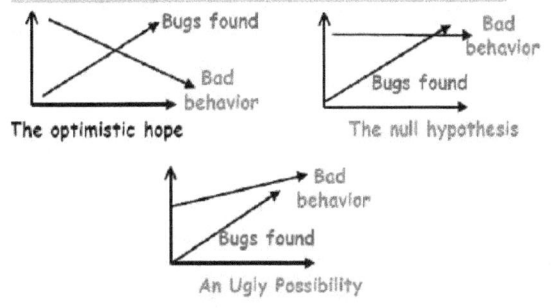

Danger: Opportunity cost.
Danger: Deterministic canary bugs to non-deterministic.

Laws of static bug finding

- Vacuous tautologies that imply trouble
  Can't find code, can't check.
  Can't compiler code, can't check.
- A nice, balancing empirical tautology
  If can find code
  AND checked system is big
  AND can compile (enough) of it
  THEN: will always find serious errors.

- A nice special case:
  Check rule never checked? Always find bugs. Otherwise immediate kneejerk: what wrong with checker???

Secure Coding Standards

Robert C. Seacord
CERT/CC
Software Engineering Institute
Carnegie Mellon University
Pittsburgh, PA 15213 USA
+1-412-228-7608
rcs@cert.org

ABSTRACT
Secure coding standards define rules and recommendations to guide the development of secure software systems. Establishing secure coding standards provides a basis for secure system development as well as a common set of criteria that can be used to measure and evaluate software development efforts and software development tools and processes. This paper describes plans by the CERT/Coordination Center at the Software Engineering Institute at Carnegie Mellon University to establish, through a coordinated community effort, a set of secure coding standards for commonly used programming languages.

Keywords
Security, Standardization, Programming languages.

1. INTRODUCTION
Society's increased dependency on networked software systems has been matched by an increase in the number of attacks aimed at these systems. These attacks—directed at governments, corporations, educational institutions, and individuals—have resulted in loss and compromise of sensitive data, system damage, lost productivity, and financial loss [19].

Software vulnerability reports continue to grow at an alarming rate [1] and a significant number of them result in technical alerts [2]. To address this growing threat, the introduction of software vulnerabilities during software development and ongoing maintenance must be significantly curtailed.

An essential element of secure software development is well documented and enforceable coding standards. Coding standards encourage programmers to follow a uniform set of rules and guidelines determined by the requirements of the project and organization, rather than by the programmer's familiarity or preference. Once established, these standards can be used as a metric to evaluate source code (using manual or automated processes) to determine compliance with the standard.

There are numerous available sources, both online and in print, containing coding guidelines, best practices, suggestions, and tips. For example, the following books have been published containing C/C++ programming languages rules and guidelines:

- C++ Coding Standards: 101 Rules, Guidelines, and Best Practices [21]
- Effective C++ : 55 Specific Ways to Improve Your Programs and Designs (3rd Edition) [10]
- More Effective C++: 35 New Ways to Improve Your Programs and Designs [11]
- Effective STL: 50 Specific Ways to Improve Your Use of the Standard Template Library [12]
- C++ Programming Guidelines [16]
- C Programming Guidelines [17]

Industry-specific standards such as the Motor Industry Software Reliability Association (MISRA) Guidelines for the use of the C language in critical systems [13] have also been published. Additionally, many companies have internal coding standards. An example of a publicly released coding standard is the Joint Strike Fighter Air Vehicle C++ Coding Standards [9].

Many online sources of coding practices and coding rules also exist, including the Build Security In web site [4] sponsored by the U.S. Department of Homeland Security (DHS) National Cyber Security Division. The SAMATE Reference Dataset (SRD), maintained by NIST [15], provides a set of programs with known weaknesses in code, design, or architecture that can lead to exploitable vulnerabilities. The Common Weaknesses Enumeration (CWE), maintained by MITRE, is a dictionary of known security weaknesses in code, design, and architecture that can lead to exploitable vulnerabilities [14].

With all these sources of information, it might seem that a secure coding standard for these languages would be unnecessary. However, none of these sources provides a prescriptive set of secure coding standards that can be uniformly applied in the development of a software system. This conclusion is reinforced by the Secure Software Assurance Common Body of Knowledge [18] published by the U.S. Department of Homeland Security, which laments the "lack of public standards as such for secure programming."

2. SCOPE
At one extreme, a secure coding standard can be developed for a particular release of a compiler from a particular vendor. At the

---

Permission to make digital or hard copies of all or part of this work for personal or classroom use is granted without fee provided that copies are not made or distributed for profit or commercial advantage and that copies bear this notice and the full citation on the first page.

other extreme, the standards can be designed to be not only compiler independent but also language independent.

A coding standard for a particular compiler release has the largest possible benefit to the smallest group of users. Targeting a particular compiler allows for the definition of rules and guidelines that deal specifically with the peculiarities of that implementation, including defects in the implementation and non-standard extensions. At the other extreme, a language-independent coding standard has the least possible benefit to the largest possible group of users, as the rules and guidelines specified at this level of abstraction are largely notional.

The secure coding standards proposed by CERT are based on documented standard language versions as defined by official or *de facto* standards organizations. For example, secure coding standards are planned for the following languages:

- C programming language (ISO/IEC 9899:1999) [5]
- C++ programming language (ISO/IEC 9899:1999) [6]
- Sun Microsystems' Java2 Platform Standard Edition 5.0 API Specification [20]
- C# programming language (ISO/IEC 23270:2003) [7]

Applicable technical corrigenda and documented language extensions such as the ISO/IEC TR 24731 extensions to the C library [8] will also be considered.

The scope allows specific guidance to be provided to broad classes of users. Programming language standards, like those created by ISO/IEC, are primarily intended for compiler implementers. Secure coding standards are ancillary documents that provide rules and guidance directly to developers who program languages defined by these standards.

3. GOALS

The goal of each coding standard is to define a set of rules that are necessary (but not sufficient) to ensure the security of software systems developing in the respective programming languages.

A secure coding standard consists of *rules* and *recommendations*. Coding practices are defined to be rules when all of the following conditions are met

1. Violation of the coding practice will result in a security flaw that may result in an exploitable vulnerability.
2. There is an enumerable set of exceptional conditions (or no such conditions) where violating the coding practice is necessary to ensure the correct behavior for the program.
3. Conformance to the coding practice can be verified.

Rules must be followed to claim compliance with a standard unless an exceptional condition exists. If an exceptional condition is claimed, the exception must correspond to a pre-defined exceptional condition and the application of this exception must be documented in the source code.

Recommendations are guidelines or suggestions. Coding practices are defined to be recommendations when all of the following conditions are met

1. Application of the coding practice is likely to improve system security.

2. One or more of the requirements necessary for a coding practice to be considered a rule cannot be met.

Compliance with recommendations is not necessary to claim compliance with a coding standard. It is possible, however, to claim compliance with one or more verifiable guidelines. The set of recommendations that a particular development effort adopts depends on the security requirements of the final software product. Projects with high-security requirements can dedicate more resources to security, and are thus likely to adopt a larger set of recommendations.

4. DEVELOPMENT PROCESS

The development of a secure coding standard for any programming language is a difficult undertaking that requires significant community involvement. To produce standards of the highest possible quality, CERT is implementing the following development process:

1. Rules and recommendations for a coding standard are solicited from the communities involved in the development and application of each programming language, including the formal or de facto standard bodies responsible for the documented standard.

2. These rules and recommendations are edited by senior members of the CERT technical staff for content and style and placed in the Secure Coding area of CERT web site for comment and review [3].

3. The user community may then comment on the publically posted content using threaded discussions and other communication tools. Once a consensus develops that the rule or recommendation is appropriate and correct the final rule is incorporated into the coding standard.

Various groups, including the ISO/IEC JTC1/SC22/WG14 international standardization working group for the C programming language have expressed an interest in supporting this model.

5. USAGE

These rules may be extended with organization-specific rules. However, the rules contained in a standard must be obeyed to claim compliance with the standard.

Training may be developed to educate software professionals regarding the appropriate application of secure coding standards. After passing an examination, these trained programmers may also be certified as secure coding professionals.

Once a secure coding standard has been established, tools can be developed or modified to determine compliance with the standard. One of the conditions for a coding practice to be considered a rule is that conformance can be verified. Verification can be performed manually or automated. Manual verification can be labour intensive and error prone. Tool verification is also problematic in that the ability of a static analysis tool to detect all violations of a rule must be proven for each product release, to detect regression errors. Even with these challenges, automated validation may be the only economically scalable solution to validate conformance with the coding standard.

Software analysis tools may be certified as being able to verify compliance with the secure coding standard. Compliant software systems may be certified as compliant by a properly authorized certification body by the application of certified tools.

6. SYSTEM QUALITIES

Security is one of many system attributes that must be considered in the selection and application of a coding standard. Other attributes of interest include safety, portability, reliability, availability, maintainability, readability, and performance.

Many of these attributes are interrelated in interesting ways. For example, readability is an attribute of maintainability; both are important for limiting the introduction of defects during maintenance that could result in security flaws or reliability issues. Reliability and availability require proper resources management, which contributes also to the safety and security of the system. System attributes such as performance and security are often in conflict requiring tradeoffs to be considered.

The purpose of the secure coding standard is to promote software security. However, because of the relationship between security and other system attributes, the coding standards may provide recommendations that deal primarily with some other system attribute that also has a significant impact on security. The dual nature of these recommendations will be noted in the standard.

7. CONCLUSIONS

The development of secure coding standards is a necessary step to stem the ever-increasing threat from software vulnerabilities. Establishing secure coding standards allows for a common set of criteria that can be used to measure and evaluate software development efforts and software development tools and processes. Once established, secure coding standards can be incrementally improved, as a common understanding of existing problems and solutions allows for the development of more advanced security solutions.

8. ACKNOWLEDGMENTS

Thanks to Thomas Plum for suggesting this idea, John Benito for supporting this effort, and Hal Burch for his insights. Thanks to Jason Rafail, Jeff Gennari, Allen Householder, Chad Dougherty, and Claire Dixon for their review and thoughtful comments.

9. REFERENCES

[1] CERT/CC. See http://www.cert.org/stats/cert_stats.html for current statistics.

[2] CERT/CC. US-CERT's Technical Cyber Security Alerts. http://www.us-cert.gov/cas/techalerts/index.html

[3] CERT/CC. Secure Coding web site. http://www.cert.org/secure-coding/

[4] DHS. Build Security In web site. See https://buildsecurityin.us-cert.gov/

[5] INCITS/ISO/IEC 9899-1999. Programming Languages — C, Second Edition, 1999.

[6] INCITS/ISO/IEC 14882-2003. Programming Languages — C++, Second Edition, 2003.

[7] INCITS/ISO/IEC 23270-2003. Information technology - C# Language Specification, 2003.

[8] ISO/IEC WDTR 24731. Specification for Secure C Library Functions, 2004.

[9] Lockheed Martin. Joint Strike Fighter Air Vehicle C++ Coding Standards for the System Development and Demonstration Program. Document Number 2RDU00001 Rev C. December 2005.

[10] Meyers, Scott. Effective C++ : 55 Specific Ways to Improve Your Programs and Designs (3rd Edition). Addison-Wesley Professional. (September 2, 1997)

[11] Meyers, Scott. More Effective C++: 35 New Ways to Improve Your Programs and Designs. Addison-Wesley Professional. (December 29, 1995)

[12] Meyers, Scott. Effective STL: 50 Specific Ways to Improve Your Use of the Standard Template Library. Addison-Wesley Professional. (June 6, 2001)

[13] MISRA C: 2004 Guidelines for the use of the C language in critical systems. MIRA Limited. Warwickshire, UK. October 2004. ISBN 0 9524156 4

[14] MITRE. Common Weaknesses Enumeration (CWE). See http://cve.mitre.org/cwe/

[15] NIST. SAMATE Reference Dataset (SRD). See http://samate.nist.gov/SRD/srdFiles/

[16] Plum, Thomas. C Programming Guidelines. Plum Hall; 2nd edition (June 1989). ISBN: 0911537074.

[17] Plum, Thomas. C++ Programming. Plum Hall (November 1991) ISBN: 0911537104.

[18] Redwine, Jr. Samuel T, Editor. Secure Software Assurance: A Guide to the Common Body of Knowledge to Produce, Acquire, and Sustain Secure Software Draft Version 0.9. January 2006.

[19] Seacord, R. *Secure Coding in C and C++*. Addison-Wesley, 2005. See http://www.cert.org/books/secure-coding for news and errata.

[20] Sun Microsystems. Java2 Platform Standard Edition 5.0 API Specification, 2004. http://java.sun.com/j2se/1.5.0/docs/api/index.html

[21] Sutter, Herb. Alexandrescu, Andrei. C++ Coding Standards: 101 Rules, Guidelines, and Best Practices. Addison-Wesley Professional (October 25, 2004). ISBN: 0321113586.

Language Design for Verification

Rod Chapman, Peter Amey

Praxis High Integrity Systems
20 Manvers Street,
Bath BA1 1PX
UK.
sparkinfo@praxis-his.com

Abstract. This position paper offers a brief summary of our experience in designing high-integrity programming languages and their associated verification tools, particularly in relation to SPARK—an annotated, pure subset of Ada95. We also consider the fundamental features of Ada that make SPARK possible in the first place, and address the question why we can't do "SPARK for X" (where X is one of today's current favorite languages). Our experience suggests that simple, small languages can offer a depth and soundness of static verification that is unachievable with today's standard languages.

1 Design goals for a program verification system

A programming language and verification system that aim to be suitable for high-integrity systems might have the following design goals:
- Soundness – the system must not give a false-negative result. This is the case where the tool says "Your program has no bugs" when actually it does – generally considered to be a bad thing.
- Completeness – the system should issue as few false-positive results (aka "false alarms") as possible. Too many such false alarms rapidly lead users to ignore the results of a tool, or to (perhaps more importantly) ignore the one really serious issue buried in a torrent of warnings.
- Depth – the verification system should be able to verify useful and non-trivial properties of our programs.
- Efficiency – the system must be fast enough to enable constructive and interactive use. If it takes all night to verify anything useful, then no-one will use it! Ideally, the system should be so fast as to wean programmers away from the lure of compilation and test.
- Composition – "separate verification" (somewhat akin to "separate compilation") must be possible. Addition of new program units must not invalidate the verification of existing units.
- Expressive Power – the language must be large enough for use on industrial-scale projects. (It's easy to meet the first five goals for a toy language that no-one else uses…)

These six goals are in a subtle balance, and the mix that can be achieved crucially depends on the programming language under analysis, for example:
- Soundness and efficiency are often traded. In C, C++ and Ada, for example, expression evaluation order is unspecified (i.e. a compiler can choose right-to-left or left-to-right order at its own whim, and is under no obligation to be consistent or to document its behavior). A static analysis tools, for efficiency, might choose to analyze only left-to-right order. This is efficient, but possibly *unsound* if the tool's choice disagrees with that made by the compiler. The *unspecified* or *undefined* language features are a plague on the efforts of the static analysis tool, yet contemporary languages are riddled with them.
- Analysis for any interesting deep property (e.g. "does my program have any buffer overflows?") is always inherently incomplete to some extent.
- Some language features require deep analysis, such as the analysis of pointers and aliasing, which would be too slow for constructive use. Efficiency can be achieved at the expense of soundness or completeness (or both…).

2   Language design trends

Historically, programming languages were designed as experiments in either expressive power (i.e. the addition of "OO" to C to get C++) or in compiler design. Languages were principally designed by compiler writers, not by people concerned with the provision of verification tools.

From the perspective of verification, much of the development of programming languages seems to have gone the *wrong way*—the addition of features that are harder and harder to analyze, such as OO (in particular dynamic dispatch of methods), generics, templates, call-backs, threads or tasks, and so on.

Only recently have we seen the trend reversed a little—it could be argued that Java and C# represent a simplification of C++, for instance, but as we will see, many of the central problems remain.

3   High integrity languages, subset and dialects

In the field of safety-critical systems, work on this problem has been going on since the mid 1980s at least. We can identify four broad approaches:
1. Work with the "whole language". In this approach, we try to build the "best effort" verification tool for a whole, unsubsetted standard language such as C, C++, or Ada. This is attractive to the tool vendor, because it creates a broad market for the tool. The costs are in efficiency, depth, soundness and completeness of the analysis. It is also attractive to the customer, because no real change in behavior, process or discipline is needed.
2. Work with a totally new language. There have been a few attempts at this approach, including LUCOL, NewSpeak, Euclid, and Eiffel. Of these, only Eiffel has achieved any real industrial impact.

3. Work with dialects. A "dialect" is a language with the unspecified features resolved by a known compiler and/or target computer – e.g. "C as compiled by Microsoft C version X.Y.X at optimization level Z". This approach improves precision and efficiency of analysis, at the cost of "lock in" to that particular compiler and language.[1] Significant results have been achieved, though, using this approach—examples including the Microsoft Static Driver Verifier[2], Cousot's ASTREE system[3], and C0[4].
4. High-integrity (annotated) subsets. This approach attempts to design a pure subset language, based on an existing industrially accepted language, but that eliminates unspecified or undefined behavior, so that analyses are valid for *any* compiler that implements the parent language. Some languages add *annotations* to strengthen the language beyond that achievable by subsetting alone. This approach is illustrated by SPARK[1].

4 SPARK for X?

We are often asked if we could do "SPARK for X" where X is C, C++, Java or whatever. We have to enquire further what the questioner actually means by this. If they are asking "can a best-effort, retrospective analysis tool be constructed for X that perhaps uses *some* annotations to improve things" then the answer is "Yes, but that's not our business". There is a wide (and growing) crop of such tools already available.

If the question is "Can you develop a verification system for a possibly-annotated subset of X which is sound, complete, efficient, deep, constructive and expressive enough for real industrial projects?" then that's a different matter. The answer is almost unavoidably "No."

Why is this? What makes SPARK different? Why is SPARK based in Ada in the first place? On reflection, we find three very basic features of Ada (and therefore SPARK) are crucial:

- **Separation of specification (contract) from body (implementation).** Ada's "package" mechanism strictly (and physically) separates the specification and body of a program unit. This was originally intended to facilitate separate compilation and development of large programs, but it has a huge impact on the verification system. Firstly, it gives us somewhere to put the contracts for a unit, such as the global variable list, pre-condition, post-condition and so on. Secondly, when a unit P references a unit Q in SPARK, only the specification of Q is ever consulted, and all the information we need is right there where we need it. The body of Q is *never* consulted. This means the system achieves efficiency and composition of analysis. Note that such a facility has been present in nearly all "Pascal-family" languages, such as Modula-1,2, or 3, Oberon, Eiffel and so on.
- **Scalar subtypes.** This may seem a totally innocuous feature of Ada, yet it remains core to SPARK's type system and verification approach. For those unfamiliar with the concept, this gives the programmer the ability to specify a (sub-)range of values

---

[1] Ironically, this "language lock-in" problem was cited in about 1975 as one of the big issues in the "software crisis" in the DoD that led to the Steelman requirements for the language that became Ada...

for a scalar type, recognizing that the world doesn't conveniently come in signed 2-complement "int" quantities. For example:

type Engine_ID **is range** 1 .. 4;

These types allow the programmer to express their intent in terms of real-world limits and quantities. Secondly, such types are a form of specification information that can be cross-checked and used for verification. Finally, they are used by the verification system to show that a program can never raise an exception resulting from an arithmetic overflow, range violation, buffer overflow, division by zero and so on. In teaching students embedded systems programming, McCormick reports scalar types as the single most influential language feature when comparing students' work completed in C versus Ada[9].
- **Pointers (lack thereof…)** SPARK manages to get by without the explicit use of pointer types at all. Firstly, Ada gives us parameter passing "modes" that do not depend on the explicit use of pointers.[2] Secondly, constrained (i.e. size known at compile-time) arrays are first-class types in SPARK, so you can pass them around as parameters, return them from functions with no recourse to pointers at all. Thirdly, Ada gives us its "chapter 13" for low-level programming, mapping objects to particular memory locations and so on. Finally, we come to linked data structures, for which we simply use arrays and array index values as "references"—the only catch being that you need to decide how big your "heap" is in advance. The impact of all this is dramatic—aliasing analysis is trivial (and sound…) so that gets us to the point where we can implement a full-blown verification system based on Hoare-logic and theorem-proving.

So, what about "SPARK for C, C++, C#, Java etc. etc" Considering the first two points above, we find the lack of separation of spec/body and the lack of scalar subtypes in such languages to be show-stopping weaknesses. Furthermore, these are hardly difficulties that can be "subsetted away" or "annotated back in" to such languages. Finally, these languages are so pointer-centric that it seems unlikely that a usefully expressive subset could be achieved that solved the "aliasing problem" to our satisfaction. We actually attempted a design study for "SPADE C" in the early 1990s—the result was so poor in expressive power and needed so much annotation that the project was not pursued any further.

5 Future and on-going work

SPARK is very much in the "raise the ceiling" mode, trying to push the high-end of static analysis, with the (non-trivial) catch that we require users to actually learn and use an entirely new language and to have the discipline and process to use it effectively. SPARK has grown significantly over the years, adding OO support, Ravenscar tasking, modular types and so on from Ada95, without sacrificing the soundness of

---

[2] A compiler *can* use pass-by-reference mechanism, but that's its business, and can't affect the semantics of SPARK.

the verification system. We are currently working on the next major expansion of the language: the adoption of a subset of Ada's generics facility. We may even be able to pick up some of the new features of Ada2005[5] if they prove useful.

The "raise the floor" community has also made substantial progress—the current crop of "whole language" analysis tools offer a sophistication of analysis that was undreamt of a few years ago, and these are having a significant impact on a much larger group of engineers and projects than SPARK probably ever will.

There are also signs of life in the research community. The needs of the security-critical market have prompted a real renaissance in static analysis. We find (much to our amusement) that "annotations" are fashionable and embodied in systems such as ESC/Java2, Microsoft's PreFast and Spec#[6] and so on. Finally, new languages designed from scratch are making a come-back. For example, Microsoft have Sing#[7], and the Coyotos project at JHU[8] is a language (BitC), verification environment and operating system that have been developed from scratch for high-integrity applications.

References

1. Barnes, J.: High Integrity Software: The SPARK Approach to Safety and Security. Addison Wesley, April 2003. ISBN 0-321-13616-0.
2. Static Driver Verifier – Finding Driver Bugs at Compile Time. Microsoft. http://www.microsoft.com/whdc/devtools/tools/sdv.mspx
3. The ASTRÉE Static Analyzer. Patrick Cousot et al. http://www.astree.ens.fr/
4. Verisoft Consortium. The Verisoft Project. See http://www.verisoft.de/ and www.verisoft.de/.rsrc/PublikationSeite/PaulVSTTE05-final.pdf
5. Rationale for. Ada 2005. John Barnes. John Barnes Informatics. http://www.gnat.com/home/ada_answers/ada_2005
6. Righting Software. Jim Larus et al. IEEE Software, May/June 2004. http://research.microsoft.com/~larus/
7. An Overview of the Singularity Project. Galen Hunt et al. Microsoft Research Technical Report MSR-TR-2005-135. Also at http://research.microsoft.com/~larus/
8. The Coyotos Secure Operating System. Jonathan Shapiro et al. John Hopkins University. See http://www.coyotos.org/
9. Software Engineering Education: On the Right Track. John McCormick, University of Northern Iowa. CrossTalk Journal, August 2000. http://www.stsc.hill.af.mil/crossTalk/2000/08/mccormick.html

Automated Calculation of Software Behavior with Function Extraction (FX) for Trustworthy and Predictable Execution

Richard C. Linger, Stacy J. Prowell, and Mark Pleszkoch
*CERT STAR*Lab*
Software Engineering Institute
Carnegie Mellon University
Pittsburgh, PA
rlinger@sei.cmu.edu, sprowell@cert.org, mpleszko@cert.org

Abstract

*CERT STAR*Lab at the SEI is developing function extraction (FX) technology to compute the behavior of software to the maximum extent possible. FX capitalizes on a view of programs as mathematical functions or relations that illuminates methods for behavior calculation. While behavior calculation is a very difficult problem, routine availability of computed behavior could have substantial impact on software engineering in general and software assurance in particular. As a first application of function extraction technology, STAR*Lab is developing the Function Extraction for Malicious Code (FX/MC) system to analyze the behavior of malicious code expressed in the Intel instruction set. FX technology provides foundations for automated security attribute analysis, correctness verification, and component composition.*

1. Computing Software Behavior

The ever-increasing complexity of software systems places extraordinary demands on human comprehension. Traditional code reading and inspection methods are subject to human fallibility and can be overwhelmed by the sheer size of programs, and software tools generally provide only partial views of program behavior. In today's state of art, no practical means exists to answer the straightforward question of what programs do in all circumstances of use. The resulting loss of intellectual control has been a persistent problem in software development, leading to unpleasant surprises from unforeseen behavior despite best efforts. What is needed is an "all cases of behavior" view for complete analysis.

It is well understood that the problem of computing program behavior is extremely difficult; however, the substantial value of such a capability motivates a closer look at what can be achieved. CERT STAR*Lab at the SEI is developing the emerging technology of function extraction (FX), with the objective of computing the behavior of software to the maximum extent possible.

The starting point for behavior computation is a precise definition of the functional semantics of instructions in the language of interest, together with rules for their combination. Sequential logic is expressible in terms of fundamental control structures, namely, sequences, alternations, and iterations (loops), whose functional semantics define the rules of combination. Thus, a required preliminary step is automated transformation of programs under analysis into structured form based on the constructive proof of a structure theorem. This transformation creates an algebraic framework for stepwise traversal and accumulation of program behavior. For sequence structures, the rule of combination is ordinary function composition. Behavior computation for sequence structures thus requires composing individual instructions to derive their net functional effect in the procedure-free form of concurrent assignments of inputs to outputs. Behavior computation for alternation structures is carried out in terms of case analysis to derive procedure-free conditional rules that organize the effects of true and false branch operations in terms of concurrent assignments. It is fortunate that the behavior of sequence and alternation structures, which typically comprise the bulk of sequential logic, can be computed in such a straightforward manner. Because no general theory for loop behavior computation can exist, engineering solutions are being developed.

2. FX Treats Programs Like Equations

Short of an impractical expenditure of time and effort, programmers have no means to determine the full behavior of programs. Despite best efforts, programs are routinely fielded with unknown behavior that may contain errors, vulnerabilities, or malicious code. The totality of large program behavior is difficult to understand because it is distributed across a virtually infinite number of possible execution paths. Testing selects paths from this set and so cannot reveal full behavior. However, large programs are at the same time composed of a finite number of control structures, each of which makes a finite contribution to overall behavior.

The function-theoretic view focuses not on program paths, but rather on control structures and mathematical foundations for their refinement, abstraction, and verification [1]. In this view, control structures are treated as rules for mathematical functions or relations, that is, mappings from domains to ranges, no matter what subject matter they may address. In particular, function-theoretic foundations prescribe procedure-free equations that represent net effects on data of common control structures and provide a starting point for behavior extraction. These equations are expressed in terms of function composition, case analysis, and, for iteration structures, a recursive expression based on an equivalence of iteration and alternation structures. Representative equations are given below for control structures labeled P, data operations g and h, predicate p, and program function f.

The program function of a sequence control structure (P: g; h) can be given by

$$f = [P] = [g; h] = [h] \circ [g]$$

where square brackets denote the program function and "o" denotes the composition operator. That is, the program function of a sequence can be calculated by ordinary function composition of its constituent parts. The program function of an alternation control structure (P: if p then g else h endif) can be given by

$$f = [P] = [\text{if p then g else h endif}]$$
$$= ([p] = \text{true} \rightarrow [g] \mid [p] = \text{false} \rightarrow [h])$$

where | is the "or" symbol. That is, the program function is given by a case analysis of the true and false branches, and the possibility of abstracting them to a single case. The program function of a terminating iteration control structure (P: while p do g enddo) can be expressed as

$$f = [P] = [\text{while p do g enddo}]$$
$$= [\text{if p then g; while p do g enddo endif}]$$
$$= [\text{if p then g; f endif}]$$

and f must therefore satisfy

$$f = ([p] = \text{true} \rightarrow [f] \circ [g] \mid [p] = \text{false} \rightarrow I)$$

These equations define an algebra of functions that can be applied bottom up to the control structure hierarchy of a program in a stepwise function extraction process. This process propagates and preserves the net effect of control structures through successive levels of abstraction while leaving behind complexities of local computations and data not required for expressing behavior at higher levels. Additional methods are required to simplify and reduce intermediate expressions and to analyze loop operations, as well as to present behavior catalogs to users in appropriate forms.

In notional illustration, consider the following miniature sequence that operates on logical variables and the question of deriving its behavior, which is not immediately obvious ($\vee$ represents the "exclusive or" operation):

```
do
    x := x ∨ z
    z := x ∨ z
    x := x ∨ z
    y := x ∨ y
    x := x ∨ y
    y := x ∨ y
enddo
```

The behavior can be computed in a trace table that accumulates intermediate compositions to arrive at net effects the intentional variables x, y, and z (I for identity):

Operation	x	y	z
$x := x \vee z$	$X = x \vee z$	I	I
$z := x \vee z$	I	I	$z = (x \vee z) \vee z$ $= x$
$x := x \vee z$	$x = (x \vee z) \vee x$ $= z$	I	I
$y := x \vee y$	I	$y = z \vee y$	I
$x := x \vee y$	$x = z \vee (z \vee y)$ $= y$	I	I
$y := x \vee y$	I	$y = y \vee (z \vee y)$ $= z$	I

Thus, the functional behavior is given by a sequence-free concurrent assignment of initial values to final values

$$x, y, z := y, z, x$$

that is, the effect of the programmed sequence of operations is to rotate the truth values of the three variables. Such behavior computations are readily automated in the function extraction process. In this case, the calculations involve logical variables and their rules of combination, but any data types and structures can be accommodated. While the behavior of this simple sequence is defined by a concurrent assignment, the general form of behavior definitions is necessarily a non-procedural conditional concurrent assignment (CCA)

predicate $\rightarrow$
    assignment 1
    assignment 2
    ...
    assignment n

where if the predicate on program input values is true, the assignments are carried out concurrently. For larger and more complex programs, the function extraction process produces behavior catalogs containing sets of disjoint CCAs that together define program behavior for all cases.

The conditional concurrent assignment is the principal statement in the behavior expression language of function extraction.

3. FX Improves Software Comprehension

STAR*Lab has developed a proof-of-concept function extractor prototype that calculates the behavior of programs expressed in a small subset of the Java programming language and presents it to users in the form of behavior catalogs. The catalogs contain procedure-free CCAs that define the net functional effect of programs from input to output in all circumstances of use.

The prototype was employed in a rigorous experiment to compare traditional methods of program reading and inspection with FX-based methods. Twenty-six experienced programmers were divided into a control group using traditional methods and an experimental group using the FX prototype. Each group was required to answer questions dealing with comprehension and verification of three Java programs. The experiment produced the following results [2]:

- The experimental group using the FX prototype reduced the time required to derive the functional behavior of the programs by several orders of magnitude compared to the control group.
- For the most difficult program, the experimental group was about four times better at providing correct answers to the comprehension and verification questions, and required a fourth of the time to do so, a productivity improvement of a factor of 15 over the control group.
- The experimental group achieved these results with 45 minutes of instruction on use of the function extractor, compared to years of training and experience for the control group.

4. FX Impacts the Software Lifecycle

Function extraction technology can be applied to any programming language environment, and has potential to impact many aspects of the software engineering lifecycle. To better understand this impact, STAR*Lab conducted a comprehensive study with a major aerospace corporation to determine how FX could improve engineering operations in activities ranging from software specification and design to implementation and testing [3]. This study produced guidance for FX evolution from experienced software developers:

- Development of FX automation for assembly language should be a priority.
- FX automation should be developed for correctness verification of software.
- FX automation should be developed for high-level languages starting with Java.
- Research on FX automation for specification and architecture should be initiated.

5. Development of the Function Extraction for Malicious Code System

CERT STAR*Lab has initiated development of the first application of FX technology in the Function Extraction for Malicious Code (FX/MC) system [4]. The goal of FX/MC is to compute the behavior of malicious code expressed in Intel assembly language, to enable security analysts to quickly determine intruder objectives and develop countermeasures. The initial version applies a structure theorem to transform intentionally obfuscated, spaghetti-logic control flow into readable structured form, and computes the behavior of sequence and alternation structures.

In miniature illustration of FX/MC capabilities, consider the following assembly language fragment that gives the appearance of being intentionally obfuscated:

```
    xor ebx, ebx
    mov edx, dword [ebp+4*ebx+50]
    xor eax, eax
    jmp loc_800002B
loc_800000D:
    inc ebx
    sub eax, edx
    mov edx, dword [ebp+4*ebx+50]
    sub eax, edx
    jmp loc_8000028
loc_800001B:
    xor edx, edx
    sub edx, eax
    mov eax, edx
    xor edx, edx
    jmp near ptr 8000034h
loc_8000028:
    inc ebx
    jmp short loc_800001B
loc_800002B:
    sub eax, edx
    inc ebx
    mov edx, dword [ebp+4*ebx+50]
    jmp short loc_800000D
```

The first step in FX/MC processing transforms the logic to function-equivalent structured form; in this case, the arbitrary branching (jmp instructions) is eliminated and the control flow reduces to a simple sequence of operations with no jumps present:

```
do
   xor ebx, ebx
   mov edx, dword [ebp+4*ebx+50]
   xor eax, eax
   sub eax, edx
   inc ebx
   mov edx, dword [ebp+4*ebx+50]
   inc ebx
   sub eax, edx
   mov edx, dword [ebp+4*ebx+50]
   sub eax, edx
   inc ebx
   xor edx, edx
   sub edx, eax
   mov eax, edx
   xor edx, edx
enddo
```

The behavior is then computed, resulting in the following concurrent assignments to registers EAX, EBX, and EDX, and to the zero flag (ZF), sign flag (SF), parity flag (PF), carry flag (CF), overflow flag (OF), and auxiliary carry flag (AF), which Intel semantics leave as undefined (arbitrary_value_bool). The +d symbol represents a doubleword addition:

```
[ EAX := M_dword(50 +d EBP)
       +d M_dword(54 +d EBP)
       +d M_dword(58 +d EBP)
: EBX := 3
: EDX := 0
: ZF := true
: SF := false
: PF := true
: CF := false
: OF := false
: AF := arbitrary_value_bool()
]
```

Thus, the computed behavior shows that the program sums up three consecutive doublewords starting at address EBP+50 and assigns the result to EAX, behavior that is not immediately obvious from inspecting the original code.

FX/MC is a substantial development effort; simply processing the Intel instruction set requires definition of the functional semantics of over a thousand opcodes. While the target of interest is malicious code, the system will extract the behavior of any programs expressed in assembly language. The technology developed for FX/MC can also be applied to function extractor development for other languages such as Java, C, and C++.

6. FX for Automated Security Attribute Analysis

In the current state of practice, security properties of software systems are typically assessed through subjective, labor-intensive human evaluation. STAR*Lab is investigating science foundations and engineering automation for fast and precise calculation of security properties both during system development and operation. The Computational Security Attributes (CSA) project is developing technology for augmenting human analysis of security properties with automated computational analysis. The emergence of function extraction technology, unavailable to previous researchers, provides the critical first step by deriving the functional behavior of software as a starting point for security analysis.

Security attributes are often referred to as non-functional properties, but they are in fact fully functional and dependent on the execution behavior of software. Desired security attributes can themselves be specified in functional terms, permitting software to be evaluated for conformance or not through comparison with the behavior catalogs generated by the function extraction process. Thus, computational security analysis requires defining the functional behavior required to satisfy the attributes of interest.

For example, consider the non-repudiation attribute and its definition in functional terms. Non-repudiation of changes to a dataset requires ensuring that the means for authentication of changes cannot later be refuted, which can be expressed, for example, as the following fundamental behavioral requirement

- If the dataset is changed during the execution of the software, a specified variable that identifies the user making the change is always associated with the dataset.

from which specific requirements can be derived:

- User binding: There exists a trusted function to identify the user making the change to the dataset which is invoked for every data change of interest.
- Atomic operations: The user binding and the dataset change are handled as a single atomic operation within the boundary of the software of interest.
- Traceability: Every change to the dataset is preceded by a definition of the change and identification that binds the change to the user, and every change is audited.

These requirements express the non-repudiation attribute in terms of data items and constraints on their processing. The processing can be expressed in appropriate forms, for example, as logical or quantified expressions or even conditional concurrent assignments, which can be mechanically checked against the FX-generated behavior catalogs of the software of interest for conformance or not.

7. FX for Automated Correctness Verification

In functional terms, software should do what it is designed to do and nothing more. In security terms, defective software cannot be secure. These quality and trustworthiness properties are essential but often problematic in software systems. The function extraction process derives the as-built specification of software, that is, the behavior that has actually been implemented. This derived behavior can be compared to requirements and specifications to determine if the software is indeed a correct implementation. This comparison is based on a Correctness Theorem that defines conditions required for correctness [1]. In addition, FX technology prescribes effective means to create and record specifications, with the corresponding specification task itself amenable to automated support. Automated correctness verification would be especially valuable during system development, to check on the behavior of partial implementations and find and fix errors and vulnerabilities along the way. It would also permit a new level of rigor in acquisition and acceptance of systems by requiring provision of behavior catalogs for all delivered code.

8. FX for Automated Component Composition

Function extractors must provide substantial composition capabilities because behavior calculation is essentially a compositional task. Creating function extractors to compose software components in systems is thus a question of scale, not of method. Given behavior catalogs for each component and the intended structure of their interaction, the composition process requires calculating the net functional effects of the combined component behaviors. As a step in this direction, STAR*Lab has conducted research on flow structures, which provide mathematical foundations and engineering techniques for analyzing and designing component compositions to satisfy mission objectives at the network architecture level [5,6,7]. Capabilities for automated composition would provide support for construction and integration of entire systems. Significant research will be required in interface ontologies and subject-matter abstractions to augment the component composition process.

9. References

[1] Prowell, S., Trammell, C., Linger, R., and Poore, J., Cleanroom *Software Engineering: Technology and Process*, SEI Series in Software Engineering, Addison Wesley Longman, Reading, MA, 1999.

[2] Collins, R., Walton, G., Hevner, A., and Linger, R., *The CERT Function Extraction Experiment: Quantifying FX Impact on Software Comprehension and Verification*, (CMU/SEI-2005-TN-047), Software Engineering Institute, Carnegie Mellon University, Pittsburgh, PA, 2005.

[3] Hevner, A., Linger, R., Collins, R., Pleszkoch, M., Prowell, S., and Walton, G., *The Impact of Function Extraction Technology on Next-Generation Software Engineering*, (CMU/SEI-2005-TR-015), Software Engineering Institute, Carnegie Mellon University, Pittsburgh, PA, 2005.

[4] Pleszkoch M., and Linger, R., "Improving Network System Security with Function Extraction Technology for Automated Calculation of Program Behavior," *Proceedings of 37th Hawaii International Conference on System Sciences,* Hawaii, January, 2004, IEEE Computer Society Press, Los Alamitos, CA, 2004

[5] Hevner, A., Linger, R., Sobel, A., and Walton, G., "Specifying Large-Scale, Adaptive Systems with Flow-Service-Quality (FSQ) Objects," *Proceedings of the 10th OOPSLA Workshop on Behavioral Semantics,* Tampa, FL, October, 2001, ACM Press, New York, 2001.

[6] Hevner, A, Linger, R., Sobel, A., and Walton, G., "The Flow-Service-Quality Framework: Unified Engineering for Large-Scale, Adaptive Systems," *Proceedings of the 35th Annual Hawaii International Conference on System Sciences,* Hawaii, January 7-10, 2002, IEEE Computer Society Press, Los Alamitos, CA, 2002.

[7] Linger, R., Pleszkoch, M., Walton, G., and Hevner, A., *Flow-Service-Quality (FSQ) Engineering: Foundations for Network System Analysis and Design*, (CMU-SEI-2002-TN-019), Software Engineering Institute, Carnegie Mellon University, Pittsburgh, PA, 2002.

Support for Whole-Program Analysis and the Verification of the One-Definition Rule in C++

Dan Quinlan[1], Richard Vuduc[1], Thomas Panas[1], Jochen Härdtlein[2], and Andreas Sæbjørnsen[3]

[1]Center for Applied Scientific Computing, Lawrence Livermore National Laboratory,
{dquinlan,richie,panas2}@llnl.gov
[2]Department of Computer Science, University of Erlangen-Nuremberg, Germany, haerdtlein@cs.fau.de
[3]Department of Physics, University of Oslo, Norway, andsebjo@student.matnat.uio.no

Abstract

We present a compact and accurate representation of a whole-program abstract syntax tree, and use it to detect a specific security vulnerability in C++ programs known as a *One-Definition Rule* (ODR) violation. The ODR states that types and functions appearing in multiple compilation units must be defined identically. However, no current compiler can enforce ODR because doing so requires the ability to see the full application source at once; where ODR is violated, the program is incorrect. Moreover, a lack of ODR enforcement makes a program vulnerable to the so-called *VPTR exploit*, in which an object's virtual function table is replaced by malicious code. Our representation of the whole program preserves all features of the source for analysis and transformation, and permits a million-line application to fit entirely in the memory of a workstation with 1 GB of RAM.

1 Introduction

Most whole-program analyses use some form of summarization, at the loss of analysis precision, since analysis time complexity is often super-linear. The traditional unit to analyze and summarize is a procedure since it does not require the compiler to see the full source at once [33]. However, suppose we provide the compiler with a complete view of the entire program. Then, the compiler may freely choose any convenient unit regardless of procedure or module boundaries, and thereby control the size, contents, and context of the program fragment to analyze [19, 35, 24]. Such techniques permit focused and efficient analyses of customizable precision. For software security assurance, improvements in precision raise the level of assurance we can guarantee.

We describe a scalable whole-program analysis that requires the full source to verify a fundamental assumption that all C++ compilers make but no compiler checks. This assumption is the *One Definition Rule* (ODR) [4], which essentially states that a C++ program is only legal if type and function definitions appearing in multiple source files are defined identically (Section 2). Code violating ODR is not legal and may not be translated to a correct executable. Nevertheless, no compiler verifies ODR because each compiles only a subset of an entire program at one time, under *separate compilation*; as it happens, only a whole-program analysis of the full source can be used to verify ODR.

A lack of ODR enforcement enables the *VPTR exploit*, a virtual function table attack [31]. Though not yet widely used, this exploit can be implemented as a simple insider attack, particularly in collaborative or open-source projects [29] (Section 3). Its use is expected to grow as defenses against stack smashing techniques mature [28]. Checking ODR is an essential preventative measure.

We implement basic support for whole-program analysis in the form of a compact and accurate abstract-syntax tree representation of an entire program (Section 4). We can store a million-line application in the memory of a single workstation having 1 GB of RAM without losing any of the information present in the original source. We achieve memory-efficiency for C and C++ programs by merging common declarations (typically appearing in header files) that might otherwise be stored redundantly for each source file. Our representation complements existing whole-program analyses by providing a simple, high-level view of the complete source from which those analyses can be derived.

We are developing this work using ROSE, an open infrastructure for building compiler-based source-to-source analysis and transformation tools [32] (Section 4). For C and C++, ROSE fully supports all language features, preserves all source information for use in analysis, and permits arbitrarily complex source-level translation via its rewrite system. Although research in the ROSE project emphasizes performance optimization, ROSE contains many of the components common to any compiler infrastructure, and thus supports the development of general source-based analysis and transformation tools. This paper summarizes aspects of ROSE especially relevant to security analysis research (Section 5).

2 One-Definition Rule (ODR)

This section summarizes the essential features of the one-definition rule (ODR). The ODR states that templates, types, functions, and certain entities can only be defined "once," in a sense made precise in the ANSI/ISO C++ Standard [4, Sec. 3.2, pp. 23–24]. Three of the main conditions of the ODR are:

1. Within a single translation unit (a source file and its headers), there may be at most one definition of any variable, function, class type, enumeration type, or template.[1] All compilers verify this condition.

2. Within the entire program, there may only be one definition of every *non-inline* function or object; an inline function must be defined in every translation unit in which it is used, with all such definitions being identical as described in Condition 3 below. Because compilers typically process only one translation unit at a time, the C++ standard does not require that compilers check this condition.

3. Some entities, including class types, enumeration types, inline functions with external linkage, and various template entities, may be defined in more than one translation unit *provided* the definitions are "identical." The C++ standard lays out the meaning of identical precisely; one notable property is that two definitions must "consist of the same sequence of tokens" to be considered the same [4, p. 24]. We use this token-based property in our ODR checker. Like Condition 2 above, compilers typically do not or cannot verify whether multiple definitions are identical as laid out by the C++ standard.

[1]There may, however, be multiple non-defining declarations, such as function prototypes, "extern" variable declarations, forward class declarations.

Listing 1: **main.cc**–A simple program

```
int main () {
    extern void runModule (void); // Module to call
    runModule ();
    return 0;
}
```

A legal C++ program must obey the ODR. However, because the standard assumes that a compiler will see only one translation unit at a time (Condition 1), it does not require that a compiler detects violations across translation units.

The linker can partially verify ODR by detecting, for instance, multiple definitions of *non-inline* functions and global variables (Condition 2). However, *inline* function ODR violations cannot be detected; these violations require a whole-program analysis.

3 VPTR Exploit

The *VPTR exploit* replaces an object's virtual function table pointer ("VPTR") with one containing malicious code [31]. The simplest technique redefines the existing definition of an inline virtual function; since a typical compiler does not see the whole program, it cannot enforce the ODR to catch instances of this exploit. This form is most easily implemented as an insider attack, which could occur in a collaborative software development environment as demonstrated by the 2003 Linux kernel backdoor [29]. Moreover, the exploit is an instance of more general *pointer subterfuge* attacks [28].

Listings 1–3 show a program containing the vulnerability. In Listing 1 at line 3, the program executes a routine defined in an external module. That routine creates two stack-allocated objects, a and b, both of type **Derived**, at line 13 of Listing 3. The **Derived** type inherits from an abstract base class (**Base**), implements the virtual method, **Derived::run**, at line 7, and declares a 1-byte datum at line 8. However, because the run method is virtual and defined as (implicitly) inline, we must redefine the method in every translation unit in which it is used, albeit with the same definition (see Condition 3 in Section 2). If the compiler cannot enforce this condition, an attacker can re-implement the method in another translation unit to execute arbitrarily different code.

We implement a basic VPTR exploit in Listing 4. This code is a separate module that defines another malicious version of **Derived::run()** in lines 6–9. Most compilers, including GCC, assume ODR holds

Listing 2: **Base.hh**–An abstract base class

```
class Base {
public:
  virtual ~Base (void) {}
  virtual void run (void) = 0;
};
```

Listing 3: **Module.cc**–An innocuous module

```
#include "Base.hh"

// Derived class, intended to be private to this module.
class Derived : public Base {
public:
  Derived (void) { buf_[0] = 'a'; }
  void run (void) { buf_[0] = 'z'; }
  char buf_[1];
};

// Public interface to this module.
void runModule (void) {
  Derived a, b; // Two instances on the stack
  Base *pa = &a, *pb = &b;
  pb->run (); // Expect b.buf_[0] == 'z'
  pa->run (); // Expect a.buf_[0] == 'z'
}
```

Listing 4: **ViolateODR.cc**–Basic VPTR exploit

```
#include <iostream>
#include "Base.hh"

class Derived : public Base { // Class violating ODR
public:
  void run (void) {
    std :: cout << "*** Hostile takeover ***"
                << std::endl;
  }
};

Derived d; // Instantiate to get malicious 'Derived'
```

and simply choose the first one encountered at link-time. That is, when compiling with

```
g++ main.o Module.o ViolateODR.o ...
```

the compiler chooses **Derived**::run() from Listing 3, whereas in

```
g++ main.o ViolateODR.o Module.o ...
```

it chooses the implementation from Listing 4. Moreover, if the application uses shared or dynamically-loaded libraries, the malicious module need only appear first in the shared library path to be executed.

VPTR exploits have more sophisticated forms, as shown in Listing 5. This example builds on the basic exploit in Listing 4 by violating ODR and then using buffer-overrun techniques to rewrite the VPTR directly. The first step on line 15 of this alternative **Derived**::run() has the same behavior as Listing 3 at line 7, perhaps to make the code appear to behave safely. However, it then executes additional malicious code in lines 16–17.

These additional lines use the fact that a derived object often stores not just its data, but the VPTR appropriate for that object's type. For example, the a and b stack-allocated instances of **Derived** declared on line 13 of Listing 3 might appear on the stack as shown in the left-half of Figure 1. Each object has its 1-byte datum, buf_[0], plus a hidden 4-byte VPTR. When line 15 of Listing 3 invokes our malicious run(), it does so on data allocated and laid out according to the definition of **Derived** in Listing 3. Lines 16–17 of Listing 5 use platform-specific knowledge of how this data is laid out to write beyond the bounds of the data and, in this case, into the VPTR of the next object on the stack, as illustrated in the right-half of Figure 1. The new VPTR is simply the address of a compatible VPTR for the **Attacker** class defined in Listing 5. The **Attacker** class contains another malicious implementation of run(). This additional form of the VPTR exploit builds on the ODR violation, so checking ODR helps defend against VPTR exploits generally.

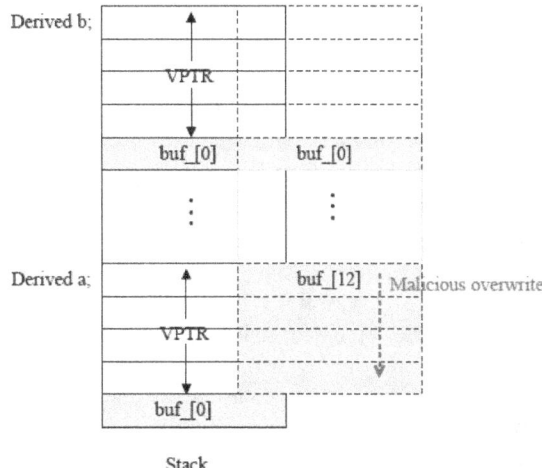

Figure 1: **VPTR exploit**. The attacker implements the alternative version of **Derived**::run() shown in Listing 5 such that executing b.run() overwrites a's VPTR.

Listing 5: **Attacker.cc**–A malicious module

```cpp
#include <iostream>
#include "Base.hh"

class Attacker : public Base { // More malicious code
public: void run (void) {
    std :: cout << "*** vtable overwritten! ***"
                << std::endl;
    // ... Do malicious things here ...
  }
};

class Derived : public Base { // Class violating ODR
public:
  void run (void) {
    buf_[0] = 'z'; // Looks normal, but see below...
    Attacker x; // Instantiate to get a vtable to inject
    *((unsigned *)(buf_+12)) = *((const unsigned *)(&x));
  }
  char buf_[16]; // Buffer used to overwrite vtable
} d; // Instantiate to get malicious 'Derived'
```

4 A Whole-Program Analysis to Detect ODR Violations

Whole-program analysis is typically implemented using procedure summaries or by embedding information into the object files to use whole-program context at link-time. Summarization is necessary to mitigate the impact of super-linear analysis time costs, and procedures are a convenient unit. However, a compiler or analysis tool should be free to analyze any useful, arbitrarily partitioned unit of the program, given a complete and accurate view of program context [35, 36, 24]. This need motivates our whole-program abstract syntax tree representation.

Below, we describe this representation as implemented in ROSE, an open and extensible infrastructure for building customized source-to-source analysis and transformation tools. A typical ROSE-based tool looks like a traditional compiler, with a front-end that generates an object-oriented abstract syntax tree (AST), a "mid-end" performing custom analyses and/or transformations to the AST, and a back-end to unparse the possibly modified AST back into source code. This section outlines recent work to extend the AST to allow the creation of a single, compact AST for the entire program. ODR violations appear during the construction of such a whole-program AST. For more information on the complete ROSE architecture, including features relevant to security analysis, see Section 5.

4.1 Overview of the whole-program representation and ODR test

ROSE's intermediate representation (IR), SAGEIII, stores all high-level information from the source code, sufficient to reproduce the original source code completely. The IR is space-efficient by design since we target large-scale physics applications of 100 KLOC per file and up. Current workstation memory capacities are also quite large (commonly 2–4 GB and greater), and so are better able to support representations of applications consisting of hundreds of files. For greater space savings, we share parts of the AST (subtrees) that are determined to be identical. This test for matching subtrees is where we check ODR, since identical definitions will by construction be shared across multiple files in the AST.

ROSE routinely compiles million-line applications file-by-file. In round numbers, these applications have on the order of 1000 files containing 75K lines contributed from header files and 1K lines of source code in the source file. The effective 76K lines of code generates an AST with about 500K IR nodes. Merging the 75K lines over each of the 1000 files thus saves 75 million lines of code from being represented redundantly in the AST. Using a 250 KLOC program, we have estimated that a million-line application will fit into approximately 400 MB of memory after merging header files. The AST holding the million-line application can also be saved to and loaded from disk using a custom ROSE-specific binary file format; on current single-processor desktop machines, writing one of these binary files to disk takes roughly 30 sec and reading less than a minute. Simple traversals of the whole AST in memory take only a few seconds. Thus, the representation is compact and efficient to operate on once constructed.

We perform the *ODR test* by unparsing candidate subtrees and verifying an exact match. Since ROSE can optionally normalize whitespace and optionally strip comments and preprocessor directives, simple string matching verifies token-by-token equivalence of the original code as required by ODR.

4.2 Whole-program AST construction

Given the ASTs from separate translation units, we merge them as follows:

1. Build an extended *mangled name map*
   The matching process is based on an extended form of name mangling that is common for handling C++ types, variables, and functions. In short, we traverse all declarations in the global scope and all namespace scopes, and for

each declaration, generate and store each declaration's unique name (*i.e.*, extended mangled name) into an STL map. The map's key is the unique name, and its value is a pointer to the associated IR node. (There are a number of details that we omit for simplicity.)

2. Build a *replacement multimap*
   The AST is traversed a second time to match the unique names generated from declarations with keys in the *mangled name map*. All matches are recorded, and a map of pairs of IR node pointers is generated (the IR node of the match and the IR node associated with the matching key from the *mangled name map*). The *ODR test* (see end of Section 4.1) is applied and must pass to be included in the *replacement multimap*.

3. Fixup AST and build the *subtree delete list*
   Using the *replacement mutimap* we traverse the AST again and find all pointers to IR nodes and using the pointer to the IR node as a key we look them up in the *replacement multimap*. If found, we replace the pointer to the key with the pointer to the value obtained from the multimap using the key and the replaced pointer value is added to the *subtree delete list*. All IR nodes that are shared via the merge process are explicitly marked as shared in the AST.

4. Delete redundant subtrees
   To save space we cannot remove redundant subtrees in the modified AST; we iterate over the delete list (which points to redundant subtrees) and remove all the nodes in each subtree.

4.3 Merged AST example

Figure 2 (top) shows the AST for the three source files shown in Listings 1, 3, and 5, with AST subtrees colored by file. The ASTs from the files are not shared. Figure 2 (middle) shows the AST after the merge process, here the diamond shaped IR nodes of the AST indicate that those IR nodes are shared. To be shared, the declaration at the root of the subtrees had to generate the same internal name (in C++ this includes standard name mangling plus a number of other language specific details) *and* the subtrees had to pass the ODR test of equivalence. Figure 2 (bottom) shows the parts of the AST which had the same internal name, but which failed the ODR test. These pairs of subtrees represent the ODR violation that enables a successful VPTR exploit.

5 The ROSE Infrastructure

We are implementing our security analysis work within ROSE, a U.S. Department of Energy (DOE) project to develop an open-source compiler infrastructure for optimizing large-scale (1 MLOC or more) DOE applications [32]. The ROSE framework enables tool builders who do not necessarily have a compiler background to build their own source-to-source translators. The current ROSE infrastructure can process C and C++ applications, and we are extending it to support Fortran90.

ROSE provides several components to build source code analyzers and source-to-source translators. The C++ front-end generates an object-oriented abstract syntax tree (AST) as an intermediate representation. The AST preserves the high-level C++ language representation so that no information about the structure of the original application (including comments and templates) is lost. This feature permits accurate analysis and the ability to regenerate the original source from the AST. The back-end unparses the AST into source code. The ROSE tool builder creates a "mid-end" to analyze or transform the AST; ROSE assists by providing a number of mid-end components, including graph visualization tools, a predefined traversal mechanism, an attribute evaluation mechanism, transformation operators to restructure the AST, program analysis support, and a number of performance optimizing transformations. ROSE also provides support for annotations whether they be contained in pragmas, comments, or separate annotation files.

Though the traditional emphasis in the ROSE project is on performance optimization, these basic components are well-suited to building software security analysis tools. A recent position paper discusses how ROSE supports the related area of automated program testing and debugging [30].

5.1 Front-end

We use the Edison Design Group C++ front-end (EDG) [13] to parse C and C++ programs. EDG generates an AST and fully evaluates all types. We translate the EDG AST into our own object-oriented AST, SAGEIII, based on Sage II and Sage++ [7]. SAGEIII is used by the mid-end as an intermediate representation. Full template support permits all templates to be instantiated in the AST. The AST passed to the mid-end represents the program and all the included header files. SAGEIII has 240 types of IR nodes, as required to represent the original structure of the application fully.

Figure 2: (*Top*) The AST before merging Listings 1 (right-most subtree in light green), 3 (left-most subtree in red), and 5 (middle subtree in blue). (*Middle*) The AST after merging. The **Base** class definition, included by Listings 3 and 5, is shared, as indicated by the magenta subtree with double-edges between diamond-shaped nodes. (*Bottom*) The merged AST, with the two **Derived** class definitions that violate the ODR shown by the subtrees with black circular nodes.

5.2 Mid-end

The mid-end permits analysis and arbitrary restructuring of the AST. Results of program analysis are accessible from AST nodes. The AST processing mechanism computes inherited and synthesized attributes on the AST. An AST restructuring operation specifies a location in the AST where code should be inserted, deleted, or replaced. Transformation operators can be built using the AST processing mechanism with AST restructuring operations.

ROSE internally implements a number of forms of procedural and inter-procedural analysis, with much of this work in current development. ROSE currently includes support for dependence, call graph, and control flow analysis. In collaboration with academic groups, we are extending the analysis infrastructure to interface with general analysis tools, including PAG [2] OpenAnalysis [34], as well as analysis tools specifically for automated debugging and security, such as Osprey for measurement unit validation [22], MOPS for finite state machine-based temporal specification checking [9], and coverage analysis tools [12].

To support whole-program analysis, ROSE has additional mechanisms to store analysis results persistently in a database (*e.g.*, SQLite), to store ASTs in binary files, and to merge multiple ASTs from the compilation of different source files into a single AST (without losing project, file and directory structure).

ROSE also provides debugging facilities, such as AST traversals and coloring, and may be used with visualization tools to aid reverse-engineering [25].

5.3 Back-end

The back-end unparses the AST and generates C++ source code. Either all included (header) files or only source files may be unparsed; this feature is important when transforming user-defined data types, for example, when adding generated methods. Comments are attached to AST nodes (within the ROSE front-end) and unparsed by the back-end. Full template handling is included with any transformed templates output in the generated source code.

6 Related Work

Whole-program analysis has traditionally been applied in performance optimization contexts [5, 35], but has recently also been used to find bugs and detect security flaws using global dataflow analyses [6, 18, 20, 14]. Our techniques complement earlier work by providing the basic infrastructure for accurately representing the source of an entire program, from which we could implement these other analyses. In the case of C++, this representation allows us to verify compliance with ODR, an important but never fully-enforced correctness condition.

Our whole-program AST is closest in spirit to the whole-program control flow graph representation proposed by Triantafyllis, *et al.* [35]. However, we essentially unify the source itself; a whole-program CFG could be easily constructed from this representation.

Atkinson and Griswold advocate on-demand generation of any representations needed for a particular analysis [5]. By contrast, we assume the exponential trends in workstation memory capacity [1] and the need for source-to-source transformation to justify generating and storing the whole-program AST.

A number of compiler infrastructures can or do perform whole-program analyses. GCC developers are adding unified cross-module representations and precompiled header support in order to provide inter-module analysis, particularly for C programs [23, 8]. Our AST merge and file I/O mechanisms are similar in spirit, though we currently provide full support for C and C++, as well as an intermediate representation that accurately represents the source. Among other open C or C++ infrastructures [16, 3, 10] and C++ static analysis infrastructures [37, 17], our complete source-level whole-program representation is unique.

7 Conclusions and Future Work

Our basic support for whole-program analysis enables any number of security analyses with complete context. The analysis we present for checking compliance with ODR to avoid VPTR exploits is just one example; the basic mechanisms permit any number of other global analyses, including whole-program pattern matching [15], region formation [35], and hybrid static/dynamic whole-program path analyses [24], among others. We will develop analyses for additional problems in collaboration with other research groups (*e.g.*, the SAMATE project [26]).

An important issue in software security analysis is how to present analysis results to users [21]. A simple textual representation of security issues is often insufficient because it is difficult to understand the context to the problem under investigation. We are investigating this problem using flexible and unique visualization techniques [27, 25].

We show an example of a program visualization in Figure 3. The program is an 80 KLOC scien-

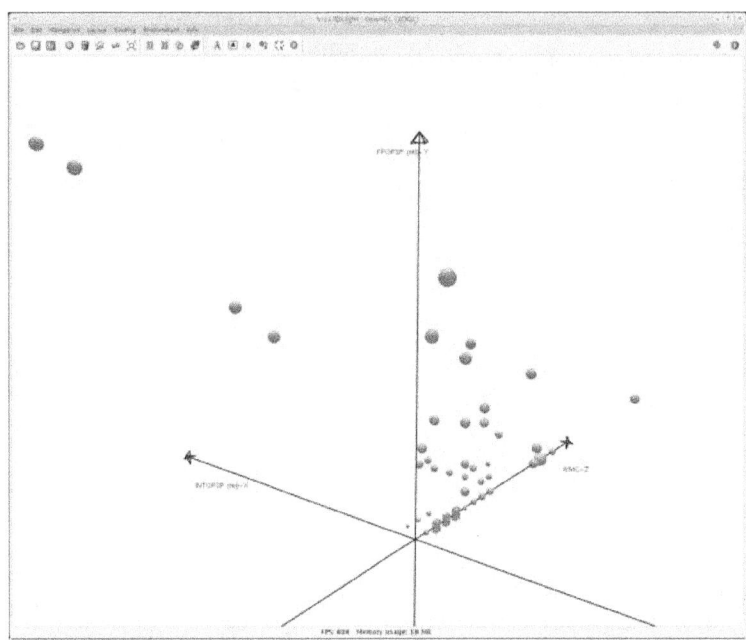

Figure 3: Visualizing security problems in source code.

tific C code, and we plot each function (shown by a sphere) according to its mathematical operations complexity, *i.e.*, the number of floating-point operations along the y-axis and the number of integer operations along the x-axis. The size of each function is equivalent to the relative size of each sphere. Furthermore, the McCabe's Cyclomatic complexity measure [11] is represented on the z-axis.

The application-specific vulnerabilities are shown by green boxes, which indicate possible program overflow problems. These vulnerable functions do not appear along either the x- or y-axis. Thus, we can infer that these vulnerable functions do not occur within the essential scientific kernels, *i.e.*, within functions that make heavy use of floating-point or integer calculations. Indeed, the problem areas for this program occur entirely within the program setup. We are pursuing this and other techniques to help users better understand security analysis results.

References

[1] International Technology Roadmap for Semiconductors, 2005. public.itrs.net.

[2] AbsInt, Inc. PAG: The Program Analysis Generator, 2006. absint.com/pag.

[3] S. P. Amarasinghe, J. M. Anderson, M. S. Lam, and C. W. Tseng. The SUIF Compiler for Scalable Parallel Machines. In *Proc. SIAM Conference on Parallel Processing for Scientific Computing*, Feb 1995.

[4] ANSI/ISO. *The C++ Standard: Incorporating Technical Corrigendum 1*, volume BS ISO/IEC 14882:2003. John Wiley and Sons, 2nd edition, 2003.

[5] D. C. Atkinson and W. G. Griswold. The design of whole-program analysis tools. In *Proc. International Conference on Software Engineering*, Berlin, Germany, March 1996.

[6] T. A. Ball and S. K. Rajamani. The SLAM project: Debugging system software via static analysis. In *Proc. Principles of Programming Languages*, January 2002.

[7] F. Bodin, P. Beckman, D. Gannon, J. Gotwals, S. Narayana, S. Srinivas, and B. Winnicka. Sage++: An Object-Oriented Toolkit and Class Library for Building Fortran and C++ Restructuring Tools. In *Proceedings. OONSKI '94*, Oregon, 1994.

[8] P. Bothner. GCC compile server. In *Proc. GCC Summit*, 2003.

[9] H. Chen, D. Dean, and D. Wagner. Model checking one million lines of C code. In *Proc. Network and Distributed System Security Symposium*, San Diego, CA, USA, February 2004.

[10] S. Chiba. Macro processing in object-oriented languages. In *TOOLS Pacific '98, Technology of Object-Oriented Languages and Systems*, 1998.

[11] S. Chidamber and C. Kemerer. A metrics suite for object-oriented design. *IEEE Transactions on Software Engineering*, 20(6), 1994.

[12] O. Edelstein, E. Farchi, E. Goldin, Y. Nir, G. Ratsaby, and S. Ur. Framework for testing multi-threaded Java programs. *Concurrency and Computation: Practice and Experience*, 15(3–5):485–499, 2003.

[13] Edison Design Group. EDG front-end. edg.com.

[14] D. Engler and M. Musuvathi. Static analysis versus software model checking for bug finding. In *Proc.International Conference on Verification, Model Checking, and Abstract Interpretation*, Venice, Italy, 2004.

[15] E. Farchi and B. R. Harrington. Assisting the code review process using simple pattern recognition. In *Proc. IBM Verification Conference*, Haifa, Israel, November 2005.

[16] F. S. Foundation. GNU Compiler Collection, 2005. gcc.gnu.org.

[17] D. Gregor and S. Schupp. Making the usage of STL safe. In J. Gibbons and J. Jeuring, editors, *Generic Programming, IFIP TC2/WG2.1 Working Conference on Generic Programming*, volume 243 of IFIP Conference Proceedings, pages 127–140. Kluwer, July 2002.

[18] S. Z. Guyer, E. D. Berger, and C. Lin. Detecting errors with configurable whole-program dataflow analysis. In *Proc. Conference on Programming Language Design and Implementation*, Berlin, Germany, 2002.

[19] R. E. Hank, W. mei W. Hwu, and B. R. Rau. Region-based compilation: An introduction and motivation. November 1995.

[20] D. L. Heine and M. S. Lam. A practical flow-sensitive and context-sensitive C and C++ memory leak detector. In *Proc. Conference on Programming Language Design and Implementation*, pages 168–181, June 2003.

[21] D. Hovemeyer and W. Pugh. Finding bugs is easy. *SIGPLAN Notices (Proceedings of Onward! at OOPSLA 2004)*, December 2004.

[22] L. Jiang and Z. Su. Osprey: A practical type system for validating the correctness of measurement units in C programs. In *Proc. International Conference on Software Engineering*, Shanghai, China, May 2006.

[23] G. Keating. Inter-module analysis in GCC. In *Proc. GCC Developers' Summit*, Ottowa, Canada, June 2005.

[24] J. R. Larus. Whole program paths. In *Proc. Conference on Programming Language Design and Implementation*, Atlanta, GA, USA, May 1999.

[25] W. Löwe and T. Panas. Rapid construction of software comprehension tools. *Intl. Journal of Software Engineering and Knowledge Engineering: Special Issue on Maturing the Practice of Software Artefacts Comprehension*, 12(54), 2005.

[26] National Institute of Standards and Technology. SAMATE–Software Assurance Metrics and Tool Evaluation, 2006. samate.nist.gov.

[27] T. Panas. *A framework for reverse engineering*. PhD thesis, December 2005.

[28] J. Pincus and B. Baker. Beyond stack smashing: Recent advances in exploiting buffer overruns. *IEEE Security and Privacy*, August 2004.

[29] K. Poulsen. Thwarted linux backdoor hints at smarter hacks, November 2003. securityfocus.com/news/7388.

[30] D. Quinlan, S. Ur, and R. Vuduc. An extensible open-source compiler infrastructure for testing. In *Proc. IBM Haifa Verification Conference*, volume LNCS 3875, pages 116–133, Haifa, Israel, November 2005.

[31] Rix. Smashing C++ VPTRs. *Phrack*, May 2000. phrack.org/show.php?p=56&a=8.

[32] M. Schordan and D. Quinlan. A source-to-source architecture for user-defined optimizations. In *Proc. Joint Modular Languages Conference*, 2003.

[33] M. Sharir and A. Pnueli. *Two approaches to interprocedural data flow analysis*, pages 189–234. 1981.

[34] M. M. Strout, J. Mellor-Crummey, and P. D. Hovland. Representation-independent program analysis. In *Proc. ACM SIGPLAN-SIGSOFT Workshop on Program Analysis for Software Tools and Engineering*, September 2005.

[35] S. Triantafyllis, M. J. Bridges, E. Raman, G. Ottoni, and D. I. August. A framework for unrestricted whole-program optimization. In *Proc. Conference on Programming Language Design and Implementation*, Ottowa, Canada, June 2006.

[36] T. Way, B. Breech, and L. Pollock. Region formation analysis with demand-driven inlining for region-based optimization. In *Proc. Conference on Parallel Architectures and Compilation Techniques*, pages 24–33, Philadelphia, PA, USA, September 2000.

[37] D. Wilkerson. OINK: A collection of composable C++ static analysis tools, 2005. freshmeat.net/projects/oink.

Towards the Industrial Scale Development of Custom Static Analyzers

Anton Ermolitski, Eric Bush, Allen Goldberg, Klaus Havelund, Doug Smith, Arnaud Venet

Kestrel Technology

3260 Hillview Real #210

Los Altos, CA 94022

{anton,ericbush,goldberg,havelund,smith,arnaud}@kestreltechnology.com

http://www.kestreltechnology.com

Abstract—This paper presents a high level overview of a technology called CodeHawk whose purpose is to support verification of software properties. Today's commercially available static analysis tools identify potential runtime and vulnerability problems based on properties described in the semantics of the programming language. While CodeHawk will detect those classes of problems, it is distinguished by the user's ability to generate high performance static analyzers for the verification of application-specific properties. Today's static analyzers may also trade off assurance and flexibility for speed in handling very large code sets. Our goal with CodeHawk is to handle industrial sized code sets with the highest speed in the industry among those offering 100% verification assurance. CodeHawk's customizability opens up additional uses of the core technology beyond detecting runtime or vulnerability exposures. In this paper we describe one such use, namely static analysis in support of optimized dynamic analysis.

I. INTRODUCTION

In this paper we present our approach to static analysis of large software systems using a platform enabling the rapid development of custom static analyzers: CodeHawk. Unlike some static analysis approaches that are optimized to identify bugs, but not prove the absence of bugs, our objective is to achieve full code coverage so that there are no false negatives with respect to a set of well-defined properties. This is appropriate for high assurance systems, particularly those that must pass a rigorous certification process. In particular CodeHawk can prove properties of a C program's memory accesses that are sufficient for 100% assurance of the absence of buffer overflow errors. Insuring there are no false negatives together with an acceptably low rate of false positives raises a challenging scaling problem. Our approach to achieving scalability is to customize the analysis to the application domain, and to use algorithms engineered for high performance.

CodeHawk is a component of a larger system that combines static analysis with dynamic analysis. Dynamic analysis refers to monitoring the execution of a program for conformance with a set of properties. Static and dynamic analysis interact in two ways. First, static analysis can either establish that a property holds, establish that it does not, or fail to come to any conclusion. Dynamic checks may be inserted in the code to assist this process. Second, checking of dynamic properties may be optimized by static analysis. Within our framework dynamic properties are complex temporal properties expressed in a rich specification notation and the validity of such a property may depend on establishing related *sub-properties* at many different program points. Static analysis may verify these sub-properties.

The remainder of this paper is organized as follows. The next section overviews abstract interpretation, the theory on which CodeHawk is based. This is followed by an overview of CodeHawk. The next section describes how domain-specific properties are incorporated into CodeHawk through motivating examples. Then we discuss dynamic analysis and its integration with static analysis. The final section states conclusions.

II. ABSTRACT INTERPRETATION

Static analysis is a generic term encompassing a variety of techniques that vary greatly in scope and nature (type checking, coding style analysis, model checking, dataflow analysis, statistical pattern inference, pointer analysis, etc.). Abstract Interpretation [5], [6] is a theoretical framework enabling the systematic construction of *sound* static analyzers. By soundness, we mean that *all* possible execution paths are taken into account in the analysis. Hence, the properties of the program discovered by such an analyzer are guaranteed to hold in any configuration of the program. Formal verification of program properties can thus be achieved by Abstract Interpretation. A precise description of Abstract Interpretation is beyond the scope of this paper. We rather give the main intuitions underpinning the theory.

The behavior of a program is described by the set of its execution traces under all possible inputs. Execution traces can be formally described using a mathematical modeling technique called *operational semantics* [3]. Abstract Interpretation allows us to build a finite machine-computable model of the operational semantics of a program using two tools: *partitioning* and *abstraction*. Partitioning consists of grouping program configurations into a finite number of disjoint sets as illustrated in Fig. 1. For example, we can partition program configurations with respect to program control points, i.e., two configurations are in the same partition iff they reach the same

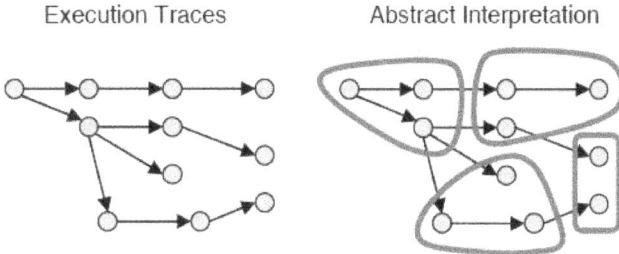

Fig 1 Partitioning of program configurations

statement in the program. Abstraction consists of defining a single finite representation of all configurations in a partition. For example, if all the program variables are integer-valued, a possible abstraction consists of assigning an interval to each variable that contains all possible values of the variable in any configuration of the partition [ ].

The abstraction process may cause the representation to denote program configurations that never occur in real executions. For example, if we have two configurations in a partition where i = 2 and i = 4, variable i is represented by the interval [2, 4], denoting the spurious configuration where i = 3. This explains the existence of *false positives* in program verification by Abstract Interpretation. A property may very well hold for all program executions, however the static analyzer cannot verify this is true, because it is violated for spurious configurations resulting from the abstraction employed. Note that we cannot have *false negatives*, i.e., a property cannot deemed true by the analyzer, even though it is violated in some executions. This is because *all* program configurations are covered by the abstraction.

Without entering in too much detail, we will just say that Abstract Interpretation provides a methodology and a collection of techniques that allow us to construct an *abstract semantic model* $\mathcal{M}$ of the program, that is a machine-representable structure representing the program dynamics on the abstract partitioning of configurations. The abstract semantic model is usually defined by induction on the syntax of the program and can be automatically generated by a proper translator. The envelope of $\mathcal{M}$, denoting the set of all possible configurations of the program, can be computed iteratively using well-studied fixpoint algorithms [ ]. This structure can then be used to conduct automatic verification of the desired program properties.

We illustrate the abstract interpretation process on a small example. Consider the following piece of C code that initializes an array of double-precision floating point numbers:

```
1:   double a[10];
2:   int i;
3:
4:   for(i = 0; i < 10; i++) {
5:     a[i] = 1.0;
6:   }
7:   a[i] = 3.0;
```

Now, assume that we are interested in assessing the correctness of all array accesses. In the example, this translates into verifying the property $0 \leq i < 10$ at lines 5 and 7. The abstract semantic model defines the level of abstraction at which the analysis algorithms will operate. For example, one can choose to ignore all information stored in data structures. This makes sense for applications like embedded systems where the control structure is essentially driven by local variables, as described in [10]. This abstraction may be inappropriate for other families of programs. Once the abstract semantic model has been determined, abstract interpretation algorithms compute an envelope of all possible values for the program variables. If we choose an abstraction of numerical variables based on intervals, the analysis will automatically infer that the range of variable i is $[0, 9]$ at program point 5, and $[10, 10]$ at program point 7. Then, the computed ranges are used to check the safety properties for array access.

III. CODEHAWK TECHNOLOGY

The Abstract Interpretation approach to static analysis looks attractive, but it presents major hurdles. Building a static analyzer based on Abstract Interpretation is a complex engineering task that can require substantial domain expertise. Designing the abstract semantic model $\mathcal{M}$ and writing the translator that takes the program text and produces $\mathcal{M}$ is the most time-consuming part of the process. oreover, the abstract semantic model is specially designed to support the verification of a small number of program properties (usually one). Scaling to large code-bases has been proven possible by tailoring the abstract semantic model toward the particular structure of the software analyzed [10], [ ]. All these factors lead to large, complex, monolithic static analyzers that are able to deal only with a handful of program properties. This approach is impractical for all but a few critical applications, and then only those blessed with a large   budget.

The purpose of the CodeHawk technology precisely consists of bringing the Abstract Interpretation approach to a practical production level. CodeHawk is built on top of the Specware [ ] formal specification environment developed by estrel Institute. The main capability offered by CodeHawk is that of building a fully functional and efficient static analyzer by assembling components drawn from a library of predefined abstractions. The Specware environment is particularly supportive of that activity. In particular, the code of the whole analysis engine can be automatically generated from the specification of the analyzer. A static analyzer checking for a certain class of properties and tailored for a specific class of software can be rapidly specified and generated using CodeHawk.

CodeHawk's precursor, C S [10] is a static array-bound checker tailored for NASA's flight mission software. It can scale up to half a million lines of C code, with a false positive rate $< 20\%$. C S is written in C and has a monolithic architecture. odifying the tool in order to have it analyze specific constructs of flight software more precisely is a complex and time consuming task. We found that the remaining 20% false positives were essentially due to array bounds transmitted between threads through message queues. odifying the abstract semantic model in order to track this information precisely was not difficult in theory, but the impact on the implementation was enormous. This basically stopped us from further specializing the analyzer. CodeHawk aims at simplifying this specialization process by generating analyzers that have a flexible and tunable architecture.

Building an abstract semantic model from scratch is facilitated by CodeHawk, but still remains the job of an expert. We are currently working toward a specification environment built on top of CodeHawk that offers the capability of specifying custom program properties to verify and generate the corresponding analyzer. This specification environment will provide a high-level interface to CodeHawk that is accessible to the non-specialist and enable the construction of static analyzers for a broad spectrum of properties. The SA ATE database [ ] will provide the basis for studying the specification language.

I . STATIC ANA SIS FRO NU ERICA SPECIFICATIONS

In this section we illustrate the concept of a specification environment men- tioned above on two examples: a string copy function and a communication application using nonces . These examples rely on the core capability currently implemented in CodeHawk: the analysis of numerical computations. They show analyzers for vulnerable use of the programmming language itself, resp. an application-specific property.

   u er er o iolations

Consider the function test defined below:

```
void test(char *str){
    char buf[10];
```
```
    memccpy(buf, str, 0, 10);
    printf("results: %s\n", buf);
}
```

which is an extract of example 000-001-31 in [ ]. The function takes as argument a string and prints it out, although in an unnecessarily complicated, and subsequently unsafe, manner, that embodies a potential for a buffer overflow. The function declares an array buf of size 10. This array is then filled up with the text string. This is done by a call of the function:

```
void *memccpy(void *s1, const void *s2,
              int c, size_t n);
```

the description of which is:

> memccpy copies bytes from memory area s into s1, stopping after the first occurrence of c has been copied, or after n bytes have been copied, whichever comes first.

The problem occurs when strlen(str) (the length of the string) is bigger than or equal to the size of the array it is copied to (here 10), since in this case a final '0' is not copied into buf, and hence if buf is now used as a string the end of this string will not be clearly marked. We want to enforce the policy that the length of the copied string is strictly less than the size of the array. Then are we sure that a final '0' is copied in.

In order to detect such an error statically, a specialized algorithm can be programmed that performs a numerical abstraction of the program and analyzes this abstraction with respect to a desired property. In this specific case the abstraction keeps track of sizes of arrays and sizes of strings, and the specification states that any call of memccpy should copy a string with a smaller size than the size of the target array. The alternative to hard coding a static analyzer for this specific problem is to apply our generic approach and synthesize a static analyzer from an abstraction specification and a property specification stating a property to be checked over the abstraction.

The *propert speci cation* now states the desired property, i.e., that calls of memccpy copy fitting strings:

```
check NoBufferOverflow =
    always(memccpy(arr, str, 0, N)
           -> size(str) < size(arr))
```

Of course this property can also be checked dynamically during program execution, and this might be a solution in case the property cannot be checked statically.

   once epetition iolations

The above example illustrated the detection of runtime errors in the form of buffer overflows. The following example illustrates a security problem concerned with uniqueness of authentification keys called *nonces*. Nonces are used for example in authentication protocols as a means of preventing replay attacks. A nonce is a "number used once". That is, the creator of the nonce should insure that it has not been used in previous runs of the protocol and that it is not guessable by an attacker.

Typically, randomly generated numbers are used as nonces. The SA ATE database [ ] includes test cases (for example example test case 000-000-05) asserting that nonces should be used for the present occasion and only once. Here we consider how static and dynamic analysis can be used to assure correct uses on nonces. We assume a protocol is implemented by a collection of procedure calls, and that if a step in the protocol requires a nonce it is a parameter to a specific procedure `sendNonce`.

The abstraction specification would in this case define an abstract state that maps each nonce to an integer indicating how many times it has been used. It will also state how this abstract state is updated as a result of execution of program statements. The property specification will state that the integer associated to a nonce should never go beyond 1. Static analysis can also be used to check that the source of the value of the nonce parameter is a random variable library function.

However, if that cannot be statically validated, dynamic analysis can check that the nonce parameter is distinct at each invocation. This property can be expressed in our EA LE monitoring language [1] as:

```
monitor NonceOnlyOnce =
   always(sendNonce(x) -> NonceNotSeen(x))

rule NonceNotSeen(int x) =
   previously(sendNonce(y) -> x != y)
```

. Co binin Static and D na ic Ana sis

Above it was mentioned that properties can be specified in a formal specification language and then checked statically. In case the static analysis cannot demonstrate the property, the whole property, or the part of the property that cannot be checked statically, can be checked dynamically during program execution using runtime monitoring techniques. A different way of thinking about this is to regard static analysis as a technique to optimize runtime monitoring: given a property to be monitored during execution, optimize the monitoring as much as possible in order to minimize the impact on execution time and memory consumption. In reality these are two views of the same problem, but from different perspectives.

These ideas can be brought even further by observing a current trend within Aspect Oriented Programming (AOP): the extension of pointcut languages with tracecuts (predicates on execution traces). In a traditional AOP language such as Aspect an aspect contains advices of the form: "when this *piece o code is encountered*, execute this other piece of code ". With *tracecuts* it is possible to state properties even more succinctly: "when this *temporal propert is true about t e e ecution trace*, execute this other piece of code " Such a framework can furthermore be supported by static analysis in the sense that static analysis statically attempts to determine when the tracecuts are true in the program and hence the new code can be inserted. In case it cannot be determined, monitors must be inserted in the code, which trigger the new code when reaching specific states.

The ODE system currently under development at estrel Technology combines static and dynamic analysis in such an AOP framework with tracecuts, in ODE called policies. ODE focuses on (1) a policy language based on state machines for expressing system safety and information assurance constraints, ( ) static analysis mechanisms for detecting policy applicability in a program, and (3) enforcement mechanisms and associated assurance arguments and evidence. An overarching objective is to lower the cost of producing certified software.

 ODE uses fast static analysis algorithms provided by CodeHawk to match each policy against the program. The engineer can specify whether to check a policy or enforce a policy. For each location in the program where static analysis determines that a policy applies, ODE either checks that it holds (generating a diagnostic message when it fails to hold) or automatically generates enforcement code for insertion at that location. ODE outputs a program that is consistent with the original program and that is guaranteed to satisfy the enforced policy.

I. Conc usion

Static analysis for 100% verification of runtime, safety and security properties, is important. But to be practical, it must satisfy two requirements. First, it must scale to application code sizes used in industry. Second, it must support verification of properties that include those better defined in terms of the application's objectives, in addition to today's focus on those defined in terms of a programming language's usage. We have introduced a technology platform called CodeHawk that can meet those requirements.

References

[1] H Barringer, A oldberg, Havelund, and Sen Rule-Based Runtime erification In roceedings o T nternational   con erence   , volume 3 of   Springer, anuary 00
[ ] F Bourdoncle Efficient chaotic iteration strategies with widenings ecture otes in omputer cience, 35, 1 3
[3] P Cousot Semantic foundations of program analysis In S S uchnick and N D ones, editors, rogram lo   nal sis   eor and pplications, chapter 10, pages 303 3   Prentice-Hall, Inc , Englewood Cliffs, 1
[ ] P Cousot and R Cousot Static determination of dynamic properties of programs In roceedings o t e econd nternational imposium on rogramming, pages 106 130 Dunod, Paris, France, 1 6

[5] P. Cousot and R. Cousot. Abstract interpretation: a unified lattice model for static analysis of programs by construction or approximation of fixpoints. In *Proceedings of the 4th Symposium on Principles of Programming Languages*, pages 238–353, 1977.

[6] P. Cousot and R. Cousot. Systematic design of program analysis frameworks. In *Conference Record of the Sixth Annual ACM Symposium on Principles of Programming Languages*, pages 269–282. ACM Press, New York, NY, 1979.

[7] P. Cousot, R. Cousot, J. Feret, L. Mauborgne, A. Miné, D. Monniaux, and X. Rival. The ASTREÉ Analyser. In *Proceedings of the European Symposium on Programming ESOP'05*, volume 3444 of *Lecture Notes in Computer Science*, pages 21–30, 2005.

[8] Kestrel. Specware System and documentation, 2003. http://www.specware.org/.

[9] NIST SAMATE Reference Dataset Project. Software Assurance Metrics and Tool Evaluation, NIST. http://samate.nist.gov/SRD/srdFiles/index.php

[10] A. Venet and G. Brat. Precise and efficient static array bound checking for large embedded C programs. In *Proceedings of the International Conference on Programming Language Design and Implementation*, pages 231–242, 2004.

Verification Tools for Software Security Bugs

Frédéric Michaud   Frédéric Painchaud
Defence Research and Development Canada – Valcartier
2459 Pie-XI Blvd North
Québec, QC
Canada G3J 1X5

June 29th, 2006

Abstract

We investigated errors and vulnerabilities that emerge from software defects in C/C++ and Java programs. This allowed us to create a meaningful testbench in order to evaluate best-of-breed automatic source code verification tools. Our results show that current static tools cannot significantly reduce the risk associated with confidential data processing in a military context. Dynamic tools should be used in conjunction in order to provide the necessary assurance level.

1 Introduction

Developing reliable and secure software has become a very challenging task, mainly because of the unmanageable complexity of the software systems we build today. Software flaws have many causes but our observations show that they mostly come from two broad sources: design (e.g., a backdoor) and implementation (e.g., a buffer overflow).

To address these problems, our research group at DRDC Valcartier first worked on design issues. A prototype of a UML design verifier was built [1]. Our approach was successful, but we faced two difficulties: specifying interesting security properties at the design level and scalability of the verification process. Building on this experience, we studied design patterns for the implementation of security mechanisms [3]. The output was a security design pattern catalog that can help software architects choose mature and proven designs instead of constantly trying to reinvent the wheel [4].

This paper addresses the implementation issues. We have evaluated automatic source code verifiers that search for program sanity and security bugs. After section 2 that specifies the context of our study, section 3 defines the terminology that we use. Then, section 4 gives the major language design shortcomings that make C/C++ programs so prone to security problems. Finally, sections 5 and 6 present an overview of the evaluated tools and the results of this evaluation, respectively.

2 Context

The assurance level required for executing applications depends on their execution context. Our context is military, in which confidential data is processed by sensitive applications running on widespread operating systems, such as Windows and Linux, and programmed in C/C++ and Java.

Our primary goal was to get rid of common security problems using automated source code verification tools for C++ and Java. To do so, we first investigated errors and vulnerabilities emerging from software defects. This allowed us to create meaningful tests in order to evaluate the detection performance and usability of these tools.

3 Defects, Errors, and Vulnerabilities

In our investigation of common software security problems, we observed that most of them do not come from the failure of security mechanisms but from failures at a lower level, which we call *program sanity problems*. Security mechanisms ensure high level properties, such as confidentiality, integrity, and availability, and are mostly related to design. Access control frameworks, intrusion prevention systems, and firewalls are all examples of security mechanisms. Program sanity problems are related to protected memory, valid control and data flow, and correct management of resources like memory, files, and network connections. Because these problems are many-sorted, a terminology is necessary to classify them.

The following definitions are based on [2]. An *error* is closely related to the execution of a program. It occurs when the behavior of a program diverges from "what it should be", from its specification. A *defect* lies in the code, it is a set of program instructions that causes an error. It can also be the lack of something, such as the lack of data validation. Finally, a *vulnerability* is a defect that causes an error that can be triggered by a malicious user to corrupt program execution.

We focused on defects, errors, and vulnerabilities that can have an impact on security. To be as general as possible, we wanted them to be application-independent. We defined 25 kinds of defects (6 categories), 5 errors, and 3 vulnerabilities, as shown in figure 1.

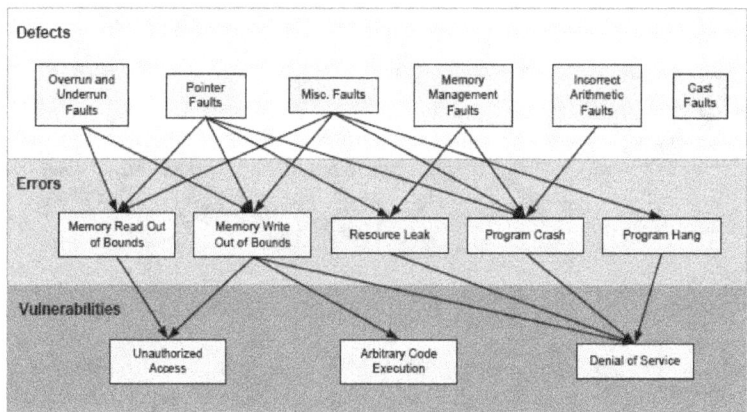

Figure 1: Defects, Errors, and Vulnerabilities

3.1 Defects

Most defects are not always "on"; they will not always generate errors for every execution of the program. Complex conditions have to be met for the error to happen and input values play an important role. Furthermore, many defects are composite and cannot be attributed to only one program instruction.

The following is a list of all defects we used to create our tests.

Memory Management Faults: Problems related to memory allocation, deallocation, and copy from one buffer to another.

1. Reading freed memory
2. Underallocated memory for a given type
3. Call of free() with an invalid pointer
4. Incorrect C++ array deletion
5. Call of memcpy() with overlapping memory regions
6. Reading uninitialized variables
7. Omitting to call non-virtual destructor of derived class

Overrun and Underrun Faults: Problems related to the overrun or underrun of an array or a C++ iterator.

1. Overrun or underrun of an array
2. Dereferencing a C++ iterator that is past the end

3. Dereferencing an erased C++ iterator
4. Incorrect size parameter to a buffer function
5. Using negative array index or size
6. Reading a string of arbitrary length without limit
7. Reading a non null-terminated string

Pointer Faults: Problems related to incorrect pointer usage.

1. Return of a pointer to a local variable
2. Incorrect pointer arithmetic
3. Dereferencing a null pointer
4. Losing resource reference

Incorrect Arithmetic Faults: Problems related to incorrect arithmetic computations.

1. Division by zero
2. Integer overflow or underflow
3. Bit shift bigger than integral type or negative

Cast Faults: Problems related to the incorrect cast of one type into another.

1. Integer sign lost because of implicit unsigned cast
2. Integer precision lost because of bad cast

Miscellaneous Faults: Problems that do not fit into any other category.

1. Unspecified format string
2. Endless loop

3.2 Errors

The list of possible low-level errors that can happen when a program is executed is very long. Since we had no interest in the correctness of computations with respect to specifications, we focused on errors that can interfere with correct memory management, control flow, and resource allocation.

Memory Write Out of Bounds A valid region of memory is overwritten. Impacts depend on what is overwritten, but this kind of error can lead to the most serious vulnerabilities since it can allow an attacker to modify the program state. The causes are generally bad pointer arithmetic and array walking with a bad index value.

Memory Read Out of Bounds A region of invalid memory is read. Impacts will mostly be errors in computations, but sensitive values could be read. The main causes are reading of a string not terminated by a null and array walking with a bad index.

Resource Leak A discardable resource (e.g., memory, file handle, network connection) is not returned to the available pool. Of course, impacts depend on the kind of resource. However, this will generally lead to a slowdown or crash of the resource-starved program. The main causes are losing a reference to a resource because of pointer reuse and the programmer simply forgetting to free the resource.

Program Hang The program is in an infinite loop or wait state, which generally lead to a denial of service. The main causes of this kind of error are never reaching a condition to exit a loop and threads in a deadlock state.

Program Crash An unrecoverable error condition happens and the execution of the program is stopped. Of course, this leads to a denial of service. The main causes are dereferencing an invalid pointer (e.g., page fault), an uncaught exception, and a division by zero.

3.3 Vulnerabilities

Errors in general are undesirable, but the real problem is vulnerabilities, especially remotely-exploitable ones. We observed that almost all dangerous vulnerabilities are associated with memory reads or writes out of bounds.

Denial of Service Allows an attacker to prevent users from getting appropriate service. It is usually done by creating an unrecoverable error condition or by exploiting a resource leak.

Unauthorized Access Allows an attacker to access functionalities or data without the required authorization. It is usually done by bypassing the control mechanism by modifying it in memory or by reading sensitive values in memory and using them to get access.

Arbitrary Code Execution Allows an attacker to take control of a process by redirecting its execution to a given instruction. It is usually done via a buffer overflow that overwrites a function pointer with the address of the injected code to execute. The return address on the execution stack is a frequent target, but any function pointer that will be called in the future is could work.

4 Why Are C/C++ Programs So Prone to Security Problems?

Many defects and errors are possible because of bad design choices made when C and C++ were created. These languages require too much "micro-management" of the program's behavior (e.g., memory management), are error-prone (e.g., pointer arithmetic), and induce serious consequences to seemingly benign errors (e.g., buffer overflows). A short list of the major C/C++ design shortcomings follows.

Lack of Type Safety: Type safety helps enforce the execution model by ensuring values assigned to variables are sound with respect to operations performed on them. Because of this, type-safe programs are *fail-fast*; their execution is stopped immediately when an error occurs. Non type-safe languages like C and C++ let the execution of erratic programs continue and many security exploits use this fact (e.g., buffer overflows).

Pointer Arithmetic: Gives the ability to a programmer to change the value of a pointer without restriction. It is then possible to read and write anywhere in the process memory space, which often lead to very obscure bugs. Furthermore, pointer arithmetic makes program verification a lot more difficult.

Static Buffers: Buffers in C/C++ cannot grow to accommodate data, buffer accesses are not checked for bounds, and overflows can overwrite memory.

C Lack of Robust String Type: C has no native type for character strings. Static buffers with overflow problems are used instead. Besides, the size of a string is indicated by a null character at the end. This is very fragile: if the null is not there, an overflow is likely to occur. C++ programs can use the string type in the Standard Template Library, but our observations show that this is rarely the case.

Creators of modern languages, such as Java, had these problems in mind and got rid of them. Indeed, Java is immune to C/C++ program sanity problems because runtime checks throw an exception if an error occurs (e.g., array access out of bounds). However, many program sanity checks throw *unchecked* exceptions and these are rarely caught by programmers. Many problems become denial-of-service vulnerabilities since uncaught exceptions crash the program.

5 Tools Overview

We evaluated 27 tools for C/C++ and 37 for Java. All these tools were categorized into 3 families: program conformance checkers, runtime testers, and advanced static analyzers.

Program conformance checkers perform a lightweight analysis based on syntax to find common defects. Because of this unsophisticated analysis, they perform poorly, except for a few defects that can be detected by simple syntax analysis (e.g., format string vulnerabilities). Many free tools were in this category.

Runtime testers look for errors while the program is running by instrumenting the code with various checks. This provides a fine-grained analysis with excellent scalability that can be very helpful when the program's behavior cannot be computed statically (e.g., because of values that are not known before runtime).

Advanced static analyzers work on program semantics instead of syntax. They generally use formal methods, such as abstract interpretation or model-checking, which often lead to scalability problems. The code must be compiled into a model and this is usually a lot more complex than it seems with C/C++ because of code portability problems between compilers (e.g., makefiles).

For C/C++, commercial tools are by far the best. For Java, however, there are many good free tools. Since Java is immune to most program sanity problems that plague C/C++, there are no exact equivalents to C/C++ tools in Java. The focus of Java tools is on good practices and high level (design) problems, such as deadlock detection. Since our goal was to detect program sanity problems, we focused on tools for C/C++ during our evaluation.

For our evaluation, our criteria were precision (flaws detected vs. false positives), scalability (small to large programs), coverage (inspection of every possible execution), and the quality of the diagnostic (report usefulness for problem correction).

6 Tools Evaluation

Preliminary tests showed that only 3 tools for C/C++ had the potential to help us achieve our goal: Coverity Prevent, PolySpace for C++, and Parasoft Insure++. We tested these tools in two ways. First, over real code in production that, to the best of our knowledge, worked well but was a bit buggy and then over many small ad-hoc pieces of code containing specific programming defects (synthetic tests).

Some tools detect errors (Insure++) and others, defects (Coverity and PolySpace). To be able to compare these tools, all results had to be converted to errors or defects. For synthetic tests, defects and the errors they cause were known in advance so it was easy to convert everything to defects. However, for code in production, nothing was known in advance, so we decided to use the best result as a baseline. Since Insure++ was the best performer, all results were converted to errors.

6.1 Synthetic Tests

A test framework with a C++ class for every kind of defect was created and integrated into a small, high-quality open-source application built with the Microsoft Foundation Classes (MFC) framework. Defects were called from the main() of the application, after initialization but before the program started to answer user queries. Defects that would lead to program crash or hang were deactivated for Insure++, since we wanted to run all tests in a single pass.

Applications built with MFC do not have a concrete main(). Instead, the program starts when the application object is created. This is a problem for PolySpace, which cannot handle that kind of main(). Therefore, it had to be used in a class-by-class analysis mode instead of a whole-program analysis. Our defects were thus designed to be detectable even without full inter-procedural analysis.

6.1.1 Results

The results of our synthetic tests are shown in figures 2, 3, and 4. No tool is totally complete and tries to detect every kind of defect or error. However, all together, tools detected all but four problems. There were no false positives, except for PolySpace that only had a few. Coverity and Insure++ focus more or less on the same kind of problems, but PolySpace, with its thorough analysis, was able to detect arithmetic and cast faults.

6.2 Code in Production Tests

The code used was a numerical analysis application of about 10K lines of code that had been in production for many years. The code was functional but a bit buggy and not very well designed (e.g., a "C+" design). As an example, we found many cut-and-pasted segments of code that could have been refactored into a method.

6.2.1 Results

The results are shown in table 1. We can clearly see that static analysis tools need good quality code to perform well. Furthermore, pointer arithmetic used to read from and write to complex data structures renders static analysis extremely difficult.

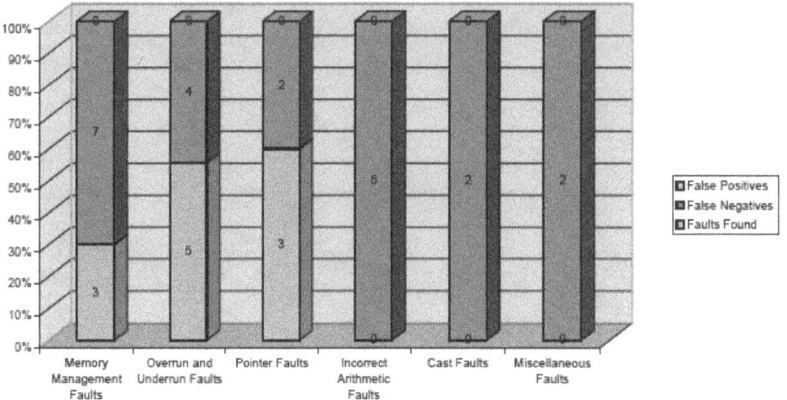

Figure 2: Coverity Prevent Results

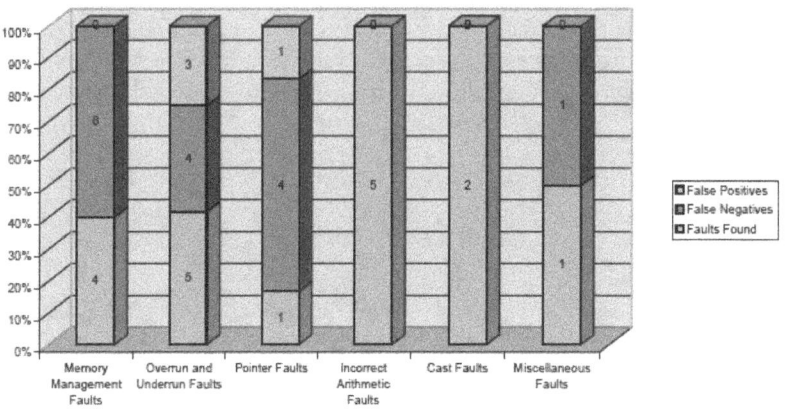

Figure 3: PolySpace for C++ Results

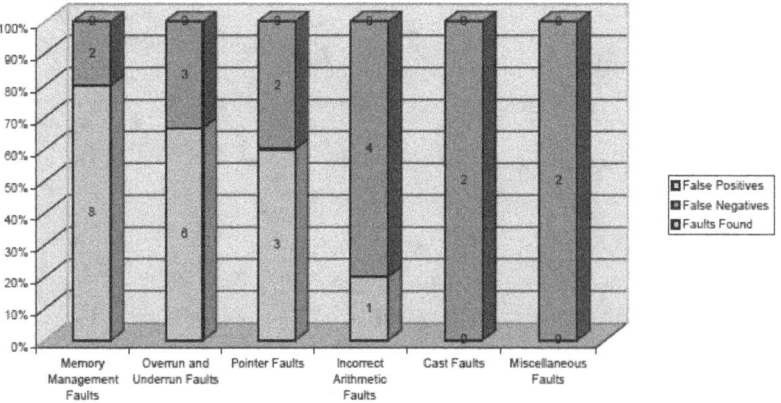

Figure 4: Parasoft Insure++ Results

Errors	Coverity	Insure++	PolySpace*
Memory Write Out of Bounds	0	42	2
Memory Read Out of Bounds	1	114	0
Resource Leak	2	10	0
Program Crash	2	0	0

Table 1: Code in Production Results
* Over 300 false positives after 16 hours of computation.

6.3 Code Portability and Makefile Problems

Static verification tools abstract programs by computing a model. From a user point of view, they can be seen as special compilers. However, making these compilers work on C/C++ code is not as easy as it sounds because C/C++ suffers from the classic code portability problem. Many C/C++ programs are compiled using makefiles, which are scripts for the *make* utility. We found that makefiles are often show-stoppers for many reasons. First, buggy makefiles are everywhere and to debug them can be a very tedious task. Furthermore, makefiles for large programs are often complex and depend on many utilities which must be configured in a very specific way. One little mistake there and nothing works. Also, when verifying large programs, one often wants to analyze only a single module or class. However, most makefiles do not offer this granularity.

Another big problem are compiler-specific extensions to C/C++. Almost all compilers support some non-standard extensions to the language and these are used a lot. The best tools have a partial support for some of them, but often, tools cannot even parse the program when these extensions are used.

When a makefile is not working properly, there is the possibility of simulating its execution. However, knowing exactly what is given to the compiler can be very hard for many reasons. First, conditional compilation using preprocessor directives is used a lot and directives often come from a mix of environment variables, configuration files, parameters to *make*, and so on. In this case, the probability of verifying a different program than the one that will be used is very high. Then, there are header file (.h) problems. Many of these files are created or moved by the makefile while it is running (e.g., .h files created by the IDL compiler on Windows). Finally, there are often many different header files with the same name, but at different locations. Knowing which one to use and when is not trivial.

We found that having the verification tool parse the program correctly is by far the biggest part of the job, and often a show-stopper unless one has unlimited time on his hands. Java is not problematic because it has no preprocessor and no conditional compilation. It has been designed to be standard and portable.

6.4 Tool Limitations and Best Usage Scenario

We found that current static verification tools suffer from what we have called the "black box problem". Indeed, for reactive applications and heterogenous systems, execution does not always take place in available application code. For instance, in reaction to a mouse click, a reactive application can start executing in kernel code to pass the event over and around the operating system. This part of its execution can rarely be analyzed and therefore, static analysis tools can hardly determine what type of data comes out of these calls. Thus, this prevents true inter-procedural analysis.

Scalability is also a problem for static tools that have to consider all possible executions (path coverage). Dynamic tools have the opposite problem: very scalable but poor coverage. However, if you consider the number of tests needed to cover all possible executions with dynamic tools, scalability is still a problem.

6.4.1 Coverity Prevent

The best usage scenario for Coverity Prevent is when the whole application needs to be analyzed and it is compiled using a working makefile. The application code size can be over 500K lines of C++ without problems. Coverity has many good points: very good integration with makefiles, uses the Edison compiler front-end that can read code that contains compiler-specific extensions from almost every big compiler in the industry (it even simulates compiler bugs!), very scalable, excellent diagnostic with execution traces that are easy to understand and very helpful to correct problems, and uses an innovative, but proprietary analysis based on statistical code analysis and heuristics. Its down sides are its primitive web interface that can be slow and the fact that it has no integration with Visual Studio projects on Windows.

6.4.2 PolySpace for C++

The best usage scenario for PolySpace for C++ is to analyze small segments of critical code in applications where runtime exceptions should never happen. The application code size must stay under 20K lines of C++. It uses a very thorough analysis based on abstract interpretation, with which it can detect runtime errors statically. It has a nice graphical interface, especially the *Viewer* module which is used to analyze the report and navigate in the source code. However, it lacks a good diagnostic because sometimes, it is impossible to understand the defect found. Moreover, it is sometimes necessary to modify the analyzed source code to have a correct model (e.g., reactive applications wait for user inputs so you have to simulate them to analyze the reactions). Its analysis stops after critical errors and the command to override this behavior is undocumented, and finally, it is slow and memory hungry, but this is expected with such a thorough analysis.

6.4.3 Parasoft Insure++

The best usage scenario for Parasoft Insure++ is to test hybrid systems based on many heterogeneous components. To consider code coverage, it should always be integrated into test case harnesses. Since Insure++ is a dynamic tool, there is no limit to the application code size and bad quality code has no effect on detection performance. Insure++ has a very good diagnostic with call stack and memory diagrams that show exactly what was overwritten. However, test cases have to be carefully specified with a good coverage strategy.

7 Conclusion

Security problems generally do not come from the failure of security mechanisms. The failure occurs at a lower level, because of program sanity problems. C/C++ are especially problematic because they enforce almost no restriction on the execution of programs and they are prone to vulnerabilities with serious consequences, such as buffer overflows. However, modern languages, such as Java, are immune to C/C++ problems and are not prone to any serious vulnerability.

Verifying C/C++ programs is a huge challenge. These languages are very difficult to analyze because of many undefined or non-standard semantics, pointer arithmetic, compiler-specific extensions to the language, etc. We have found no currently-available verification tool that can reduce the risk significantly enough for sensitive applications (please refer to section 2). We highly recommend the use of modern programming languages such as Java or C#, which nullify program sanity problems. However, if the use of C/C++ is mandatory, we recommend to restrict its usage (e.g., no pointer arithmetic, use of robust string type only, etc.) and of course to do serious test cases and use verification tools.

References

[1] Robert Charpentier and Martin Salois, *Security Modeling for C2IS in UML/OCL*, UNC, International Command & Control Research & Technology, 8th Symposium, SL 2003-067, June 2003 (18 pages).

[2] Algirdas Avizienis, Jean-Claude Laprie, Brian Randell, and Carl Landwehr, *Basic Concepts and Taxonomy of Dependable and Secure Computing*, IEEE Transactions on Dependable and Secure Computing, Vol. 1, No. 1, January-March 2004 (23 pages).

[3] Bin Chen, François Guibault, Sébastien Laflamme, Marie-Gabrielle Vallet, Yun Wang, *Secure Design Patterns: State-of-the-Art, Software Defects and Patterns Specification*, Technical Report, École Polytechnique de Montréal, December 2004 (102 pages).

[4] Bin Chen, François Guibault, Sébastien Laflamme, Marie-Gabrielle Vallet, Yun Wang, *Security Oriented Design Patterns Catalog*, Technical Report, École Polytechnique de Montréal, December 2004 (208 pages).

A framework for creating custom rules for static analysis tools

Eric Dalci   John Steven
Cigital Inc.
21351 Ridgetop Circle, Suite 400
Dulles VA 20166
(703) 404-9293
{edalci,jsteven}@cigital.com

Abstract

Code analysis tools have only a limited standard set of rule they enforce out of the box. Some code analysis tools have built in extension capabilities in the form of rule customization. Companies adopting code analysis tools gain their full benefit only through customization. This paper describes experiences with custom rules written for the Fortify Software Source Code Analyzer [1]. We explain a process that we followed to achieve reasonable accuracy and coverage. The process we describe is "tool agnostic;" we believe that it can be adopted for other code analysis tools as long as they offer a customization mechanism.

1- Introduction

Code analysis tools scan source code for implementation bugs without actually executing the source code, unlike penetration testing tools. In addition to enforcing a set of core rules out of the box some code analysis tools offer the capability to look for additional security vulnerabilities by writing custom rules. Every organization has its own specific corporate security standards. Every organization also possesses a wealth of incident data in its operations department. Both corporate standards and incident data represent essential custom rules to be created. Because of how organization-specific these data points are, tool vendors are very unlikely to create these rules as they iterate their tool. This paper introduces a testing framework for custom rules that we have created and used during a custom rule creation process for Java source code. This framework possesses three benefits. First, it improves the instances of a particular vulnerability a rule identifies, since a rule violation may appear with different code constructs within a source code—increasing true positive results. Second, the framework improves the accuracy of the rules—reducing false positives. Finally, this framework helps identify the limitations of the tool and provide new insights for the code reviews—identifying false negatives.

2- The need for custom rules

The majority of code analysis tools are based on rules that describe desirable or undesirable characteristics for a piece of software. Tool vendors will likely continue to provide updates to their set of predefined rules over time. For example the new rules may test for newly discovered software vulnerabilities. A company's own penetration testing and incident response data may be an excellent place to look for such new vulnerabilities. These breaches are pulled from one's own systems. There is automatically an applicability, feasibility, and high priority to detecting these same findings earlier.

As the tools' analysis capabilities become more mature, organizations expect more from them. Central to that expectation is customization and extension. Custom rules may be used to enforce corporate standards—which like incidents are to a certain extent necessarily different from organization to organization. Every organization seems to have a corporate standard for use and configuration of a particular set of strong cryptographic algorithms. It is unlikely that tool vendors will ever include such rules in their core set because they are not in the business of taking a stance on what is sufficient—like corporate security groups are.

3- Creating the rule, from the idea to the battle field

This section describes, step by step, our test-driven framework for creating custom code analysis rules.

Rules can be expressed at different levels of abstraction. They can be made so specific that they contain the exact code that constitutes a violation, but highly specific rules can be cumbersome and they may not even work well since the same vulnerability may manifest itself in many different ways, depending on who wrote the code.

At the opposite end of the spectrum, highly general rules may be violated by any suspicious use of a method. The typical example of this is a semantic rule flagging any occurrence of a potentially dangerous method, but without having constraints on the input and output parameters. An overly general rule typically causes an unmanageable number of false alarms (i.e., signaling security vulnerabilities where none exist). Such rules still require extensive human effort to ferret out the genuine issues and separate them from the false alarms.

In addition to that, there are different types of rules. Some rules simply look at simple semantics, and define a C function such "gets()" as unsafe. Other rules demand analysis of data flow, control flow, or configuration files. More complex code analysis tools can express rules as state machines and some can even create "partial models" of how code might execute that allow for more powerful and accurate statements about vulnerability.

Step one of rule creation involves documenting a vulnerability that can found statically. It greatly helps this first step if the performer is familiar with the custom rule creation features of the code analysis tool because there are limits to what's feasibly identified statically by each code analysis tool.

The rule can originate from multiple sources such as programmers' bug repository, corporate coding standards, incident's, published best practices, and other sources. A cryptographic rule defined in step (a) is used to illustrate our step by step process.

a. Scoping the rule.

The first step is to define and scope the rule that we want the tool to enforce. This first definition will be conceptual and not tied to a particular code construct. However the rule should be specific enough to check itself against a source code implementation.
For instance a security policy may mandate the use of strong cryptographic algorithms for secure data transmission. At the implementation level, we want to enforce the use of AES (CBC mode) and 3DES (CBC-EDE3 mode) regardless of the language, toolkit, or platform being used. Any use of unapproved algorithms would violate our rule.

b. Drafting high level axioms (optional)

The second step is to express the rule using a high level description language. Our previous cryptographic rule (described in step (a)) can be expressed with axioms that cover the different implementations that a programmer may write. The high level axioms for our rule might be as in Listing 1.

```
If [Cipher.instance]
and    (
       [used_Cipher != AES(CBC mode)] and
       [used_Cipher != 3DES(CBC-EDE3 mode)]
       )
Then
       Issue_Alarm("CipherMisused");
```

Listing 1

The rules created in this stage are just preliminary drafts; writing complete and well defined axioms will require some further exploratory work. In particular, these rules will need to be revisited after writing a first set of test cases. In fact, it may be difficult or impossible to write any axioms at all without having some test cases on hand already. In such cases, the first step may have to be omitted entirely.

c. Packaging of the test cases

To test a code analysis rule, we use code fragments which either contain rule violations (to test detection ability) or correct code (to test for false alarms). The test cases need to be organized consistently. We package the test cases within an Abstract Class or an Interface containing the java methods illustrated in *Listing 2*.

```
        void trueNegativeExamples();
        void truePositiveExamples();
        void falsePositiveExamples();
        void falseNegativeExamples();
```

Listing 2

The method `trueNegativeExamples()` will host the true negatives test cases. The method `truePositiveExamples()` will host the true positive test cases. Before the first round of testing, the content of these two first methods are hypothetical. For example, when testing a rule that scans for unauthorized cryptographic methods `trueNegativeExamples()` might contain uses of authorized cryptographic algorithms, which should not generate any violations. At the same time, `truePositiveExamples()` might contain uses of unauthorized algorithms which should lead to violations if the rule is working correctly.. The last two methods, `falsePositiveExamples()` and `falseNegativeExamples()`, are initially empty because their content is tool dependent and therefore not predictable before a code scan. Indeed, two different code analysis tools may not report the same false positives and false negatives. Identifying false negatives can be a difficult subtle game, but it is an important one. Actually catching vulnerabilities classified in our test suite, as false negatives will require manual code review, dynamic testing, or some combination. Failing to identify false negatives means you are missing vulnerabilities present in the code.

d. Writing test cases

The next step is to write test cases which will implement the correct and incorrect way to implement the rule. If an axiom has been written in the previous step, the test cases writing will be facilitated. To illustrate this step we wrote test cases for our previous example in step (a). In Java, there are many possible source code constructs to implement the use of allowed cryptographic algorithm. Therefore we can start to list all the possible correct ways to implement the use of the permissible algorithms. In the Java Cryptographic Extension (JCE) framework [2], in order to use cryptographic algorithm we should get an instance of the Cipher Object. The following code samples in *Listing 3* are all valid implementations.

```
public void trueNegativeExamples()
{
// true negative #1
// Use of AES (CBC mode)
Cipher.getInstance("AES/CBC/PKCS5Padding");

// true negative #2
// Use of 3DES (CBC-EDE3 mode)
Cipher.getInstance("DESede/CBC/PKCS5Padding");

// true negative #3
// Use of String parameters
String cipherSpec1="DESede/CBC/PKCS5Padding";
Cipher.getInstance(cipherSpec1);

// true negative #4
// Load the algorithm name from a property file
which has an authorized algorithm
Properties p = new Properties();
p.loadFromXML(new
FileInputStream(PROPERTIES_FILE));

cipherSpec2 = p.getProperty("cipherSpec");
Cipher.getInstance(cipherSpec2);

// true negative #5
String cipherSpec3="DESede/CBC/PKCS5Padding";
if (cipherSpec1.startsWith("DES"))
{
cipherSpec3 = cipherSpec1.replaceFirst("DES",
"DESede");
}
Cipher.getInstance(cipherSpec3);

// true negative #N
// etc.
}
```

Listing 3

From a static analysis perspective (with Fortify's Source Analyzer), the previous examples are considered true negatives. The code analysis tool should not report them as findings because they are all valid implementation respecting the corporate mandate on cryptographic algorithm.

Similarly, we have to list all the possible violations of the rule that we are trying to enforce. That list will be our list of true positives, the ones that the tool should recognize as violating our authorized algorithms rule.

Writing these two lists may require imagination and experience. Most of the time programmers are thinking about the right way to program things. Almost oppositely, writing test cases requires to come up with, not strictly speaking abuse case, but data (in this case source code) that will cause the code analysis tool to fail. In essence, we are stress-testing the tool. For instance, the use of an unauthorized algorithm would violate the rule as illustrated by the following code *Listing 4*.

```
private String cipherSpec1;

void init()
{
//unauthorized algorithm
cipherSpec1 = "DES/CBC/PKCS5Padding";
}
...
public void truePositiveExamples()
{
String cipherSpec2 = "AES/ECB/PKCS5Padding";

// true positive #1
// interprocedural
Cipher.getInstance(cipherSpec1);

// true positive #2
Cipher.getInstance(cipherSpec2);

// true positive #3
// concatenating Strings
StringBuffer cipherSpec3 = new
StringBuffer("IDEA");
cipherSpec3.append("/CBC/ISO10126Padding");
```

```
Cipher.getInstance(cipherSpec3.toString());

// true positive #4
// from a system property which has an
// unauthorized algorithm value.
cipherSpec3 = System.getProperty("cipherSpec");

// true positive #N
// etc.
}
```

Listing 4

The goal of having multiple test cases with similar effects is to try to cover all the different implementation variants. It bears clarifying: here we are talking about variations in syntax that might trick the code analysis tool rather than purely alternative implementations of a code construct. For instance the true positive test case #1 and #2 have the same result, their Cipher Object take an unauthorized algorithm as parameter, but the parameter passing is done differently. Specifically, one tests the tool's interprocedural parameter modeling.

Some of the test cases are intentionally too complex for the tool to recognize as true positives or true negatives, but they represent control or data flow that might occur in real application's source code in a less contrived form. For instance true positive #4 takes an environment variable which has an unauthorized algorithm as value. Static analysis tools face tremendous difficulty identifying examples like #4 because an environment variable can be resolved deterministically only at runtime. This test case can have its true negative counterpart which would take an authorized algorithm as environment variable, but again it is unlikely for the tool to be accurate unless the tool's user can provide it hints during analysis. While some vendors' tools allow such 'hints', Fortify's product does not currently.

Maturity of test cases gradually elevates as the tester can define more complex code constructs that define the tool's limits. We did not use a quantitative scale for evaluating the complexity of the test case. Instead, we relied on several years of static analysis experience. For instance we know that some code analysis tools have no inter-procedural analysis checks. Therefore we can add a test case that hides a vulnerability using an inter-procedural call.

The list of false negatives and false positives should now permit us to write well defined axioms specifying the rule at the source code level.

e. *Writing/Revisiting the source code level axioms.*

Iterating test cases allows us to iteratively refine the accuracy of axioms that will specify what code constructs would violate our custom rule. An accurate axiom would typically describe the rule constraints so the rate of false positives is reduced. In order to expedite rule writing, we used a common grammar for axiom writing. We defined the syntax of this common grammar as pseudo code similar to the specification language through which one writes certain types of custom rules for the Fortify product to facilitate translation. But ideally we would want to have larger common grammar that could cleanly express rules that rely heavily on other analysis such as control flow, data flow, or state machine specification. The axiom corresponding to the cryptographic rule in step (a) is mapping to the code construct in *Listing 5*.

```
// true negative #1
// Use of AES (CBC mode)
Cipher.getInstance("AES/CBC/PKCS5Padding");
```

Listing 5

For our cryptographic example, the Java source code axiom would look like the following Listing 6.

```
FunctionCall:

function.name == "getInstance"
and
function.parameters.length != 0
and
function.enclosingClass.supers contains [Class
name == "javax.crypto.Cipher"]
and
function.parameters[0].type=="java.lang.String"
and
not(
arguments[0].constantValue is [String:
startsWith "AES/CBC"] or

arguments[0].constantValue is [String:
startsWith "DESede/CBC"]
)
```

Listing 6

Translation into axioms crucially maps the high level requirements of a security standard to possible implementations in a particular language's source code. It is necessary to ensure that all the rule constraints are captured properly and no constraints are lost during this translation phase.

f. *Implementing the axioms using the tool extension mechanism*

Next, the rule writer implements a rule by translating axioms into whatever form the tool-specific extension mechanism requires. Some code analysis tools use a proprietary rule description language; others use programming languages as extension mechanism (C++, python, etc.)

g. Running the static analysis tool against the test cases.

The code analysis tool should take the new custom rules as input and be run against all the true positives and true negatives that we have constructed. In our experiments, this was done in a loop, iteratively, and we used our test cases as well as code "from the wild" to collect measurements on our ability to find additional true positive results and increase a rule's accuracy through reducing false positives. We believe it is reasonable to expect a 100% increase in both measures when customizing a tool's existing core rule.

h. Analyzing the results

The test scan creates two new expected categories of test cases: the false negatives and false positives (see *Figure 1*). One of the goals of code analysis is to minimize the number of findings in those two new categories. False positives create noise and require time consuming verification. This is only tolerable if the number of false positives is low. But on the opposite we want to avoid false negatives. In this case, false negatives are true rule violations that the tool missed.

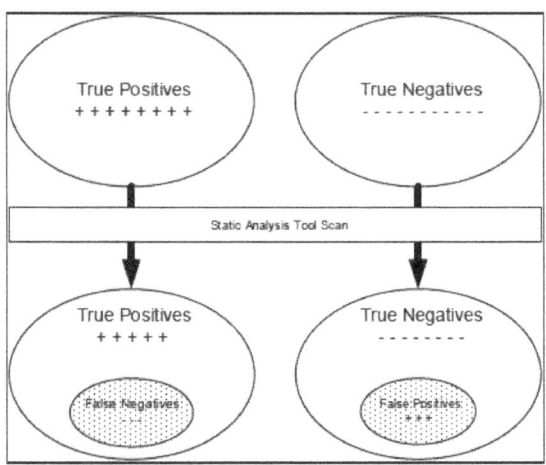

Figure 1

We can therefore reclassify the test cases according to the tool's findings. We move those test cases belonging to the new categories to their respective methods `falsePositiveExamples()` and `falseNegativeExamples()` from the true negatives and true positives methods.

At this point it is useful to try to understand what confused the tool. Why did the tool report the false positives? Why did it not catch the false negatives? Errors may be caused by the tool itself or by the implementation of the custom rule. The tool has limitations, for instance in our previous example the tool may not be capable of recognizing the value of the `String` input parameter which represents an authorized algorithm. Usually the tool user does not have much control of the tool's implementation limitations (this applies to commercial tools, where source code is not available). However, the user has control of the custom rule implementation which uses the tool's extension mechanism. Problems caused by faulty rules can be fixed, and fixing them is the purpose of the next step.

i. Feedback loop, return to step one (axioms) and stop when low false positive and false negative residuals

One of the goals of this framework is to have the static analysis tool reporting all true positives and have a low rate of false positives and false negatives. Therefore after the first iteration, the scan result may not be satisfactory. At a higher level the axioms can be incorrect and may need to be revised. The process may need multiple iterations before being accepted by the user with tolerable error levels.

It is useful to note that some code constructs are more frequently used than others, and reporting the most frequently used code constructs first will lead to the fastest results. This property is illustrated in the following Java code (*Listing 7*).

```
private String cipherSpec1;

void init()
{
//unauthorized algorithm
cipherSpec1 = "DES/CBC/PKCS5Padding";
}
...
public void truePositiveExamples()
{
String cipherSpec2 = "AES/ECB/PKCS5Padding";

// true positive #1
// interprocedural
Cipher.getInstance(cipherSpec1);

// true positive #2
Cipher.getInstance(cipherSpec2);

}
```

Listing 7

The previous *Listing 7* demonstrates that there are many ways to violate the rule. The tool is supposed to catch all true positives, but the test cases that we really care about are the first two test cases (#1 and #2). The remaining true positives test cases (*Listing 4*) are less a concern because they are unlikely to occur but we still desire that the tool covers them. We assume here that most of the programmers would use the case #1 and #2 in a real application. Trying to cover the most likely code constructs for a rule violation can be a wise choice when the possible code constructs are too numerous.

j. Integration with other process

As mentioned earlier, false negatives provide valuable insight into other security activities. False negatives represent what the tool does not find as rule violation, but should ideally. Other techniques can be used to find those false negatives depending on their severity. False negatives should feed manual code review standards, security testing efforts, and in some cases, may guide application deployment or penetration testing efforts.

Custom rules can be further tested by running them against wild code to find out if they behave as predicted in the test framework. The rules can be continually fine-tuned to achieve greater efficiency.

Conclusion

We have described a step by step test methodology that we have used to write efficient custom rules for automated software scanning. This test-driven approach has several benefits. It can expand the state of the art of static tool analysis. Identifying the undesired residual results such as false negatives and false positives can be used to improve the accuracy and coverage of existing tools. The false negatives test the tool's limits and create new technical challenges for tool providers. Being able to isolate the false negatives and positives is also crucial knowledge for the scan results reviews and manual reviews.

References

[1] Fortify Software
http://www.fortifysoftware.com/

[2] Java Cryptography Extension (JCE),
http://java.sun.com/products/jce/

[3] SAMATE NIST Project
http://samate.nist.gov

[4] McGraw G. Software Security: Building Security In, Addison-Wesley Professional, 1st Edition 2006

High fidelity static analysis for secure enterprise software requires platform knowledge

Nikolai Mansourov[1,2], Djenana Campara[1], Norman Rajala[1], Sumeet Malhotra[3]

[1] KDM Analytics,
[2] corresponding author, nick@kdmanalytics.com
[3] Unisys corp

Abstract. Static analysis methods and automatic tools that scan for security vulnerabilities offer significant advantages over manual reviews and audits. However, there are several common misconceptions about the nature and the scope of static analysis that limits its usability in certain contexts. We demonstrate that in order to perform high fidelity security analysis of entire enterprise systems the scope of static analysis needs to be increased. Lack of knowledge of the operating environment of the software may result in significant amount of false positive reports produced by a static analysis tool. This paper defines *high-fidelity static analysis*, discusses its limiting factors, the need to extend static analysis models with the representation of the operating environment of software and gives a brief overview of the Object Management Group (OMG) approach to a common representation suitable for high fidelity static analysis for security of entire enterprise systems.

1. Introduction

Automatic static analysis is positioned as an alternative to a manual code review [1]. Indeed, static analysis methods supported by automatic tools that scan for security vulnerabilities offer significant advantages over manual audits. Advantages of tool supported security analysis include *consistency* of a scanning tool (a scanning tool uses a certain formalization of security vulnerability patterns, and can be trusted to systematically explore all known possibilities), potentially broad *coverage of vulnerability patterns* (security scanner tool can use a library of vulnerability patterns created by security experts, which can in many cases exceed the expertise of an auditor), potentially broad coverage of the code, and speed of the analysis.

However, it is a well-known fact that in practice the *fidelity* of automatic static analysis is still quite low. This paper attempts to define *high-fidelity static analysis*, examines its limiting factors, and offers some insight as to why fidelity of a static analysis tool depends on the type of application. We discuss requirements for high fidelity security analysis or enterprise systems. The objective of this paper is to demonstrate that high fidelity static analysis of enterprise systems requires extensions to traditional program representations, inspired by compilers. At the end we give a brief introduction into the new Object Management Group (OMG) foundation for high-fidelity static analysis of entire enterprise systems, the Knowledge Discovery Metamodel (KDM) [2].

2. High fidelity static analysis and its limiting factors

Fidelity (of something copied or reported) is defined as truthfulness, closeness in sound, facts, color, etc. to the original[1]. Fidelity of the automatic static analysis is therefore directly related to its accuracy. Accuracy can be defined as the degree of absence of false positives reports and the soundness of the analysis, or the absence of false negative reports. On the other hand, the *power* of the automatic static analysis is directly related to the set of security vulnerability patterns, the thoroughness and the speed of the analysis. So far, this distinction allows us to separate high power but low fidelity tools (broad coverage of vulnerability patterns, but large number of false positives), and low power high fidelity tools (accurately reporting a limited number of vulnerabilities).

However, the definition of fidelity includes more than just accuracy. Fidelity of static analysis is related to how the reports that are produced by the automatic static analysis tool are close to the intended model used by developers. Therefore, fidelity is related to the representation used by the tool for analysis, and the differences of this representation from the model used by developers. In other words, fidelity of a static analysis tool is how close are the results to the ones that can be potentially produced during the manual inspection. This introduces a new distinction between an "bluntly" inaccurate report (one that is not likely to be produced during a manual review, but can be produced in large numbers by an automatic tool), and an "interesting" report, that was considered worth investigating, even if it was considered false at the end. The first is a characteristic of a low fidelity static analysis, where there is a significant disconnect between the model used by the tool, and the model used by the developers.

---

[1] Longman Dictionary of Contemporary English

Let's look at the limitations to high fidelity static analysis. Firstly, application code is not self-contained, and is seldom determined by a programming language alone. On the surface, all source code artifacts are expressed in a certain programming language. However, application code is developed for a certain *technical platform*, which significantly determines both the architecture and the control and data flow structures of the code. As the result, in most cases it is not sufficient to understand the particular programming language in order to understand the code. Consider the following simple example [3].

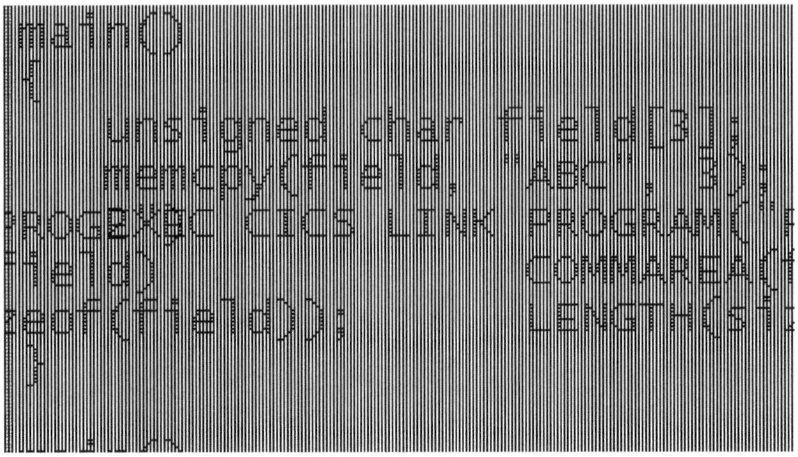

Figure 1. Example of the application code involving a technical platform

This code is written in C. The example above consists of two files, each of which defines a function called "main". The first function defines a static buffer, copies a string "ABC" into this buffer, and then uses a macro "EXEC CICS LINK" and returns. The second function defines a pointer to a buffer, then uses macro "EXEC CICS ADDRESS" and checks if the buffer contains the string "ABC". Function "main" is usually known to be the entry point into a program. However, this alone is not sufficient to understand this code. In fact, the snippet is taken from the CICS programming manual. To fully understand the behavior of this code (in order to perform high fidelity vulnerability detection), it is important to extend the model. In addition to such concepts as "file", "function", "main" function, "buffer", "usage of a macro definition" and "system call", and the capability to analyze control and data flows through statements within one procedure as well as interprocedurally, a high fidelity model for this example should also include the following concepts:

- CICS transaction
- CICS commarea
- CICS configuration

In addition to building an internal representation determined by the C language, a high fidelity static analyzer capable of processing the above example should include the capability to parse CICS configuration files, and the capability to analyze control and data flows in the extended model, where the second program is registered with CICS as a transaction with name "PROG2", and the first program invokes the second program through CICS by performing an "EXEC CICS LINK" command, and that the contents of the "field" buffer defined in the first program are made available to the second program (via the CICS commarea mechanism).

Static analysis that does not take such information into account will be low fidelity.

Related factors that may limit fidelity of static analysis (in no particular order), include the following:
- Dynamic structures (processes, threads, etc.)
- Calls via pointers
- Virtual functions
- Application frameworks
- Event-driven systems
- Reflexion
- Dynamically linked modules

Specific challenge of high fidelity static analysis is to utilize additional information about the technical platform of the software in order to complete control- and data flow paths. This is illustrated at Figure 2. It shows an *execution path* that consists of three *segments*: {1,2,3}. The path spans two *components* and involves control- and data flow, determined by the technical platform (segment 2). Segment 1 of the execution path starts at function "a", goes into function "b" of the same component, and returns to function "a". Segment 3 starts at function "c", goes into function "d", then into function "e", returns to function "d", goes into function "e", returns to "d", then to "c" and finishes. Let's assume that a certain *security vulnera*bility (for example, an unchecked access to a buffer) is detected at function "c" in component B, and the data flow required to validate this vulnerability involves function "b" in component The technical platform determines control- and data flow, by providing some function "X" (for example, an event manager,) that calls function "a" of component A, then calls function "c" of component B. This information is not explicitly present in the source code of either component. At Figure 2 the missing information is represented by dashed lines.

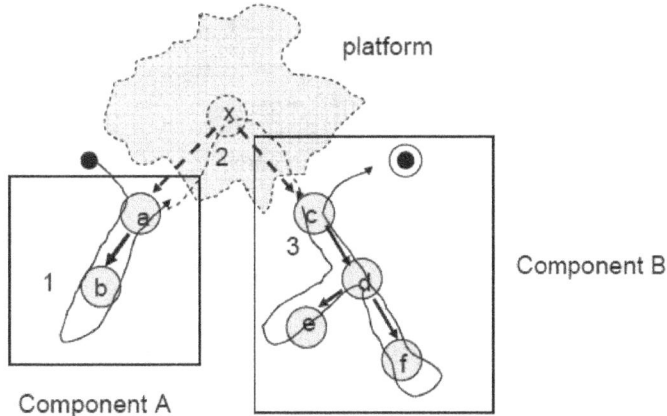

Figure 2. Platform knowledge is required to complete the execution path

Traditional representations focus on components A and B as they are determined by the programming language. Therefore they are not sufficient for high-fidelity static analysis of this example. As a side note, segments 1 and 3 in isolation involve different *binding times* than segment 2. Segments 1 and 3 involve the so-called compile- and link- time binding, while segment 2 is provided by the technical platform which involves at least deployment time binding, or initialization time binding or in some cases even a pure runtime binding. By definition, language processing tools only deal with compile time bindings. Further discussion of binding times is outside of the scope of this paper. High fidelity static analysis will increase the scope of analysis to include the knowledge of the technical platform in order to complete the execution path {1,2,3} and analyze an entire application.

There are further limitations to high-fidelity static analysis. Let's consider the differences in representations used by the tool and by developers in more detail.

The starting point is the *source code* as it is written by the application programmer. During the application building and integration this code undergoes multiple *transformations* to produce the product, which usually involves several binary deployable components. The product is then deployed (or installed). Then the product runs on the target technical platform. At runtime, it may be beneficial to distinguish between the initialization phase during which the semipermanent dynamic structures are created, and the execution phase, which may involve creation of more on-demand dynamic structures, for example, using reflexion, dynamic process creation, virtual function, callbacks, etc. Some of these techniques may be entirely user-driven. By "structures" we mean certain entities and relations between them (following the terminology recommended in [2]). Control flow and data flow relations are fully determined by these structures. Static analysis examines these structures, applies security vulnerability patterns to detect potential vulnerabilities and validates control and data flow to either confirm of reject the vulnerability. Usually, when vulnerability is confirmed, a static analysis tool is capable of identifying some sort of an execution path leading to the vulnerability. In some cases, the information produced by a static analysis tool may be sufficient to figure out the actual exploit (from the black box testing perspective).

At this level, several additional mismatches with the intended models used by developers can occur. A typical example involves the usage of the preprocessor (macrodefinitions, conditional compilation, include files). Representation for static analysis usually involves preprocessed code, however the developer works with the original code before the preprocessing. Loss of fidelity can occur in this situation, since, for example, a buffer overflow is reported only for a certain configuration (as determined by conditional compilation settings), but is not reported for other configurations. A more dramatic loss of fidelity may occur when different compilation settings are used for building the product then for static analysis.

The loss of fidelity in current tools may be caused by several common misconceptions about the scope of static analysis. Traditionally static analysis has been developed as an extension of compiler construction techniques. Therefore, typical representations used by static analysis tools focus on artifacts that are located in source files, and that are determined by a programming language. The representations may include text, lexical tokens, syntax trees, abstract syntax trees, abstract syntax graphs, and specialized program analysis representations (in the order of increasing fidelity). Analysis may involve different *scope*: local, module-level, or global. Indeed, these are the representations that have been proven useful in the area of compiler construction. The goal of the compiler is to translate these entities into some sort of binary representation. Therefore, the compiler does not involve any models of the operating environment of each module. However, the need to represent entire enterprise systems, as well as other mismatches with the models used by developers, require more information to available to the static analyzer, going beyond traditional compiler models.

3. Enterprise systems and their operating environments

In this section we will consider the landscape of applications and discuss the reasons why the same static analysis tool may provide high fidelity results in some situations and low fidelity results in others.

Let's first define the term "platform". The history of computing can be characterized by the invention of more and more powerful programming platforms for developing applications. First applications were programmed for a physical machine. Then an operating system was invented, and applications were programmed for a particular operating system. A virtual machine was invented, which added another layer in the technical platform. Further, business applications involved the usage of network systems, database management systems, and transaction systems. Further, the technical platform involved middleware and component based environments. Traditional business applications are programmed for a *technical platform* that involves most of the above elements (see Figure 3). Often, a business application will also involve a specific application framework.

Enterprise systems involve enterprise application integration (EAI) layer. A composite enterprise application uses application platform, rather than a technical platform. Modern business applications use the Service-Oriented Architecture (SOA) to define a service enablement layer. The motivation for using powerful platforms is to *close the gap* between the physical machine and the business process domain of the application. We use the term "operating environment" to refer to the entire business process platform of an enterprise application [4] (see Figure 3).

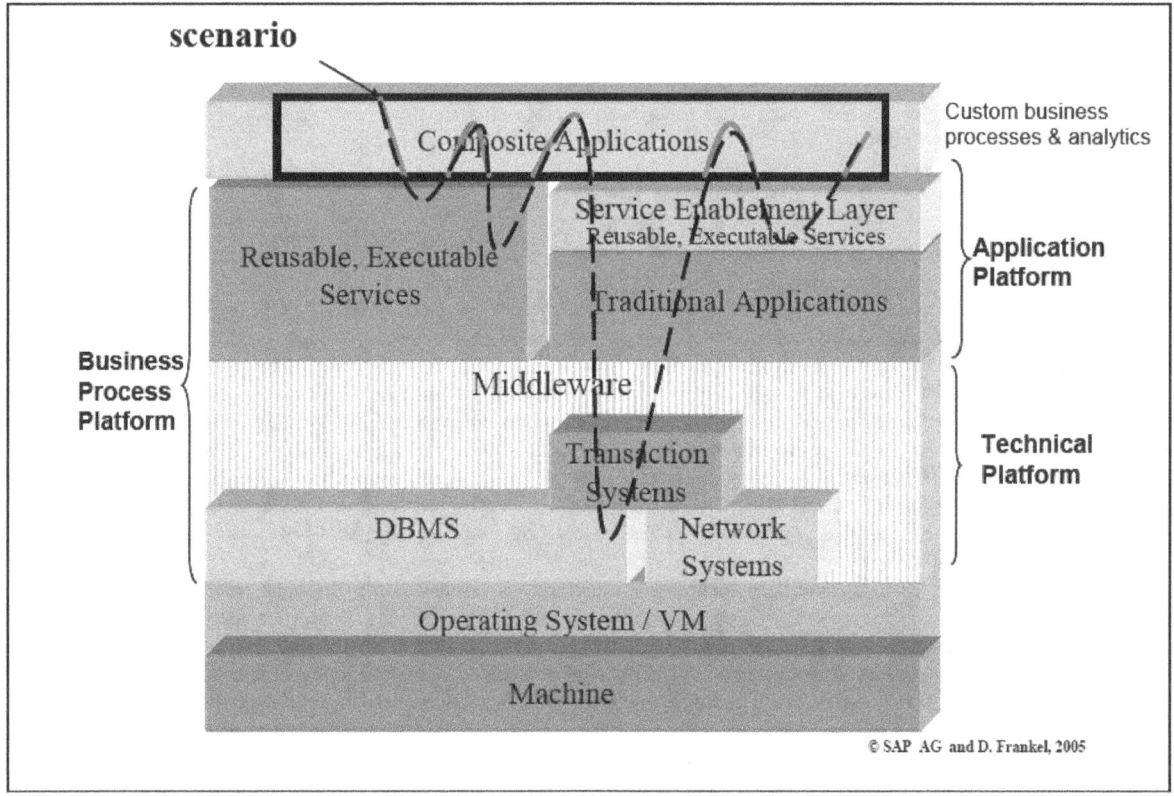

Figure 3. Operating environments of modern enterprise applications

A typical scenario through an enterprise application crosses the boundaries of composite application code, "service" applications in the application platform, and parts of the technical platform of each "service" application. In a low-fidelity code-centric approach, the explicitly visible portions of an interesting scenario may be fragmented (see Figure 3).

Enterprise systems usually integrate multiple "service" applications, which involve multiple technologies as parts of the technical platform[5]. "Service" applications usually involve multiple programming languages as well as various configuration files for describing integration, deployment, and installation. In order to perform high-fidelity analysis of enterprise applications, knowledge representations should extend well beyond specific programming languages to include the entire business process platform, while still performing control- and data flow analysis determined by the programming language statements. Pragmatically, this means that the challenge of high-fidelity static analysis of entire enterprise applications is best addressed by multiple static analysis tools, some of which specialize in extracting language specific models for particular programming languages, others specialize in extracting platform-specific information. Coordination between static analysis tools requires exchange of information based on a common representation. The rest of the paper will introduce the work done in OMG to standardize such common representation [2].

In the overall landscape of applications (for example, each box at Figure 3), some applications are more self-contained then others. For example, system programming applications, like the implementation of a DBMS, operating system code, or a network driver will only use few parts of the overall technical platform. Therefore, a static analysis tool, when applied to a system programming type of application, such as Linux kernel code, may appear high fidelity, while the same tool may show lower fidelity results for a traditional business application, and will be perceived as low fidelity for higher-end applications that use an application-specific frameworks and enterprise application integration.

4. Common representation of platform knowledge for high-fidelity static analysis

This section provides an overview of a common representation of platform knowledge that can facilitate exchange of information between different static analysis tools and can lead to high-fidelity static analysis involving entire enterprise systems.

What are the commonalities between various technical platforms?

- platform provides *resources* to application code
- platform provides *services* that are related to resources
- application code invokes platform services to *manage the life-cycle* of resources
- platform provides *component deployment mechanism*
- platform defines *control and data flow* between application components
- platform provides *error handling* across application components
- platform provides *integration* of application components

The purpose of a platform is to simplify application development by closing the gap between the application domain and the facilities that are available to application programmers. These facilities are referred to as *platform resources*. Examples of platform resources include the following: Posix File, Posix IO Stream, Posix socket, Posix Process, Posix thread, AWT widget, CICS File, CICS transaction, UNIX semaphore, UNIX shared memory segment, OS/390 VSAM file, JDBC connection, HTTP session, HTTP request, UNIX memory block, CICS commarea, COBOL file.

Usually, major platforms provide a mechanism for deploying functionality. A unit of deployment for a particular platform is further referred to as *deployment component*. Deployment component is a replaceable unit of an application. Packaging and deployment scheme usually includes configuration facility that supports assembling systems from deployment components. Configuration can occur at Deployment time, Initialization time, or at Run time. Examples of application unit include the following: DLL, shared library, COM component, Ecipse plugin, Executable,, Jar file, War file for Tomcat, SQL Stored procedure, CORBA module, EJB, JavaBean, Jakarta Struts Action, Jakarta Struts Form, Event handler, Interrupt handler.

Major platform elements support componentization by "reversing" some of the control flows. "Reversed" control flows reduce coupling between components (but not necessarily eliminate it). Deployment components are usually *plugins* into the platform. Control flow starts from inside of the platform. Platform *activates* application components through various kinds of call-back mechanisms. Knowledge of platform-specific activations is essential for understanding an enterprise software system. Examples of platform-specific activations include the following: CICS program linking, CICS transaction flow (RETURN), Unix interrupt handling, Eclipse plugin invocation, AWT event listner, CORBA method invocation, UNIX main(), WINDOWS winmain(), Servlet invocation, Jakarta Struts action: run(), Java thread run() method.

Error handling may be considered as part of inter-component control flow supported by the platform, but this is such an important aspect of application development, that it deserves a special category. Examples of platform-specific error handling includes the following: Java exception mechanism, C++ exception mechanism, COM HRESULT, CICS ABEND.

Some resources are designed to be shared between application components so that components can exchange information (data and events). Interprocess communication aspects of runtime platforms are related both to data resources, and to control-flow, as inter-component communication usually implies an indirect flow of control between components (invocation of the receiver component by the platform as the result of initiating communication by the sender component). Examples of interprocess communication mechanisms include the following: CICS commarea, CORBA message, Java RMI message, MQSeries message, HTTP request parameters, Windows event, UNIX message queue , Database notification via callback, UNIX semaphore, UNIX shared memory segment

When separation of concerns between application code and runtime platform is considered, it is important to be clear about the so-called *bindings* and various mechanisms to achieve a binding (or delay it). A binding is a common way of referring to a certain irrevocable implementation decision. Too much binding is often referred to as "hardcoding". This often results in systems that are difficult to maintain and reuse. They are often also difficult to understand. Too little binding leads to dynamic systems, where everything is resolved at run time (as late as possible). This often results in systems that are difficult to understand and error-prone. Modern platforms excel in ingenious ways to manage binding time. Usually binding is managed at deployment time. Large number of software development methodologies support efficient management of binding time, for example, portable adaptors, code generation, and model-driven architecture. Efficient management of binding time is often referred to as "platform independence".

5. The new OMG foundation for high fidelity static analysis of enterprise systems

The Object Management Group (OMG) has specified the new foundation for high-fidelity static analysis of entire enterprise representations called the Knowledge Discovery Metamodel (KDM) [2]. KDM is designed as the OMG foundation for software assurance and modernization. KDM provides a common standard way of representing and exchanging models of existing enterprise systems and their operating environments. KDM is designed as a common language and platform independent model with a powerful extension mechanism that can address language-, platform- and vendor-specific requirements. KDM is a metamodel defined in OMG Meta Object Facility (MOF). KDM specifies the common repository format, an XML-based standard exchange format (KDM XMI), and a complete API to KDM models.

KDM is aligned with a well-known architecture view approach [6]. It follows the separation of concerns principle to provide a collection of models each of which defines a common language and platform independent view of an enterprise system. The basis of KDM includes high-fidelity Code Model that represents common program elements such as procedure, variable, etc., and the Action Model that represents execution statements, and thus can be used for basic control and data flow analysis. The second level of KDM includes several models that use the primitive information captured in the Code and Action Models and represent additional information, which is not explicitly present in the source code. This level includes the Platforms & Runtime Models that provide a common way of representing the platform knowledge, according to the outline given in the previous section.

The second level of KDM also includes the following models:
- Data Model, that captures persistent data management aspects of enterprise systems,
- Build Model that captures engineering view and engineering supply chain,
- Structure Model that captures subsystems and layers of the system
- UI and Event Model that capture the user interface and presentation aspects
- Conceptual and Behaviour Models that capture domain-specific information and can be used for example for business rules mining, for representing meaningful scenarios across the system, etc.

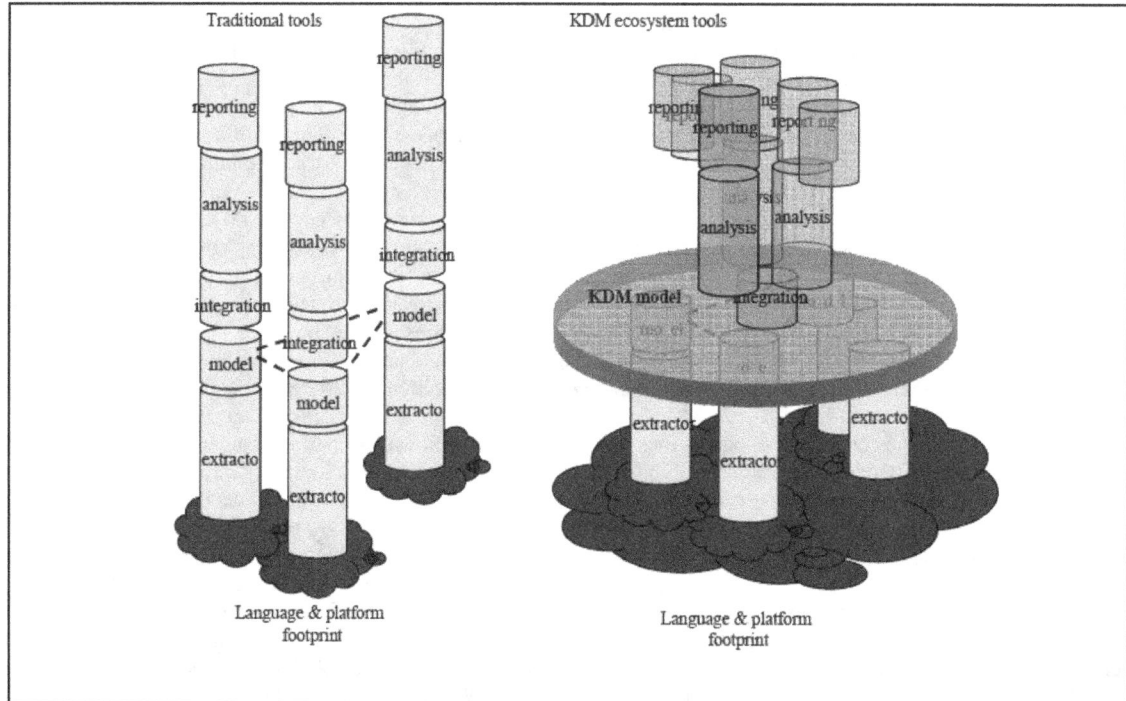

Figure 4 KDM facilitates the new ecosystem of software assurance and modernization tools

KDM leverages the collective experience of building static analysis, program understanding and modernization tools from such companies, like IBM, EDS, Unisys, ASG, and others [2]. There already exist a large industry of software tools for software assurance and modernization. However, until recently each tool is build as a stand-alone "silo", with a proprietary extractor that determines the programming language "footprint" of the tool, a proprietary analysis engine, some visualization, etc. (Figure 4, left side). Some proprietary extractors may include hardcoded knowledge of some specific platforms. Each tool includes a proprietary model with different degrees of fidelity. Currently, the exchange of information between static analysis and modernization tools is ad hoc and point-to-point, usually driven by larger software integrator companies that use these tools to perform assessment and modernization of enterprise systems (Figure 4, left side).

KDM facilitates information exchange between existing static analysis and modernization tools, as well as development of the next generation high fidelity static analysis tools. Integration between existing static analysis tools can be done by providing KDM adaptors to existing models, performing export and import of models using the standard KDM XMI representation. Next generation software assurance and modernization tools can leverage KDM API, defined in the OMG KDM standard, and the corresponding framework, SDK, and repository.

KDM facilitates the new ecosystem of software assurance and modernization tools, which emphasizes specialization in high-fidelity components that can be easily integrated into the overall framework as well as standard-based exchange of information between tools (Figure 4, right side).

6. Conclusions

High fidelity static analysis of entire enterprise systems requires significant improvements in internal representations used by traditional static analysis tools. Enterprise systems usually integrate multiple applications, involve multiple technologies that collectively comprise the business process platform. An enterprise system usually involves multiple programming languages as well as various configuration files for describing integration, deployment, and installation. In order to perform high-fidelity analysis of enterprise software, knowledge representation should extend well beyond specific programming languages to include business process platforms as well as business domain concerns, while still performing control- and data flow analysis determined by the programming language statements. Pragmatically, this means that the challenge of high-fidelity static analysis of entire enterprise software is best addressed by a consortium of multiple static analysis tools, some of which specialize in extracting language specific models for a particular programming languages, others specialize in extracting platform-specific information, yet other tools specialize in performing control and data flow analysis, and others – in security vulnerability patterns. Coordination between static analysis tools requires exchange of information based on a common representation.

The Object Management Group (OMG) has recently standardized such common representation, called Knowledge Discovery Metamodel (KDM) designed as the common language-, platform- independent and vendor-neutral foundation for high fidelity static analysis of enterprise systems. KDM includes two levels of models: the traditional compiler-like internal representation, and derivative layer, representing information that is essential for correct understanding of an entire enterprise system, but that is not explicitly available in source code. In particular, the KDM model involves a common representation of platform knowledge related to business process platforms, which is required for high fidelity static analysis of entire enterprise systems.

7. References

1. Brian Chess, Gary McGraw, Static Analysis for Security, IEEE Security & Privacy, pp 32-35, 2004
2. OMG Knowledge Discovery Metamodel, draft adopted specification, omg document amdtf/06-03-01
3. Horswill, Designing and Programming CICS Appications, O'Reilly, 2000
4. D. Frankel, Model-Driven Architecture: Applying MDA to enterprise computing, Addison-Wesley, 2004
5. Hohpe, Woolf,Enterprise Integration Patterns, Addison-Wesley, 2004
6. Hofmeister, Nord, Soni, Applied Software Architecture, Addison-Wesley, 2000

A Status Update: The Common Weaknesses Enumeration

Robert A. Martin
MITRE Corporation
202 Burlington Road
Bedford, MA 01730
1-781-271-3001

ramartin@mitre.org

Sean Barnum
Cigital, Inc.
21351 Ridgetop Circle, Suite 400
Sterling, VA 20166
1-703-404-5762

sbarnum@cigital.com

ABSTRACT
This paper is a status update on the Common Weaknesses Enumeration (CWE) initiative, one of the efforts focused on improving the utility and effectiveness of code-based security assessment technology. It is hoped that the CWE initiative will help to dramatically accelerate the use of tool-based assurance arguments in reviewing software systems for security issues.

1. INTRODUCTION
More and more organizations want assurance that the software products they acquire and develop are free of known types of security weaknesses. High quality tools and services for finding security weaknesses in code are new. The question of which tool/service is appropriate/better for a particular job is hard to answer given the lack of structure and definition in the software product assessment industry.

There are several efforts currently ongoing to begin to resolve some of these shortcomings including the Department of Homeland Security (DHS) National Cyber Security Division (NCSD) sponsored Software Assurance Metrics and Tool Evaluation (SAMATE) project [1] being led by the National Institute of Standards and Technology (NIST), and the Object Management Group (OMG) Software Assurance (SwA) Special Interest Group (SIG) [2], among others. While these efforts are well placed, timely in their objectives and will surely yield high value in the end, they both require a common description of the underlying security weaknesses that can lead to exploitable vulnerabilities in software that they are targeted to resolve. Without such a common description, many of these efforts cannot move forward in a meaningful fashion or be aligned and integrated with each other to provide strategic value.

As part of their participation in the SAMATE project, MITRE has helped lead the creation of a community of partners from industry, academia, and government to develop, review, use, and support a common weaknesses dictionary/encyclopedia that can be used by those looking for weaknesses in code, design, or architecture as well as those teaching and training software developers about the code, design, or architecture weaknesses that they should avoid due to the security problems they can have on applications, systems, and networks.

2. FIRST STEPS
The initial steps of the CWE work entailed collecting and reviewing past efforts in organizing and itemizing security weaknesses and identifying those concepts, constructs and lessons that could be used to create the CWE dictionary. Lauren Davis, from the Johns Hopkins University Applied Physics Laboratory, facilitated this work. At the same time we started establishing the foundations of a web site design to hold the materials, ideas, and documents that would come out of the CWE initiative. An important element of the CWE initiative is to be transparent to all on what we are doing, how we are doing it, and what we used to develop the CWE List. We believe this transparency is important both during the initial creation of the CWE List so that all of the participants in the CWE Community will feel comfortable with the end result and won't be hesitant about incorporating CWE into what they do. However, the transparency must also include those that will come after the CWE creation activities are complete and should be provided the opportunity to review and learn about how the CWE List was created. To this end we will be making sure that copies of all of the source documents of publicly available information used in creating CWE List are available on the web site [3].

3. PRIMING THE PUMP
To start the creation of the CWE List we brought together as much public content as possible, using three primary sources:

- the Preliminary List of Vulnerability Examples for Researchers (PLOVER) collection [4] which identified over 300 weakness types created by determining the root issues behind 1,400 of the vulnerabilities in Common Vulnerabilities and Exposures (CVE) List [5];
- the Comprehensive, Lightweight Application Security Process (CLASP) from Secure Software. which yielded over 90 weakness concepts [6], and
- the issues contained in Fortify's Seven Pernicious Kingdoms papers, which contributed over 110 weakness concepts [7]

Working from these collections as well as those contained in the other thirteen information sources listed on the CWE web site "Sources" page we developed the current draft of the CWE List, which entails almost 500 separate weaknesses.

The CWE List content is provided in several formats and will have additional formats and views into its contents added as the CWE initiative proceeds. Currently one pane of the main CWE page contains an expanding/contracting hierarchical "taxonometric" view along with an alphabetic dictionary pane. The end items in the hierarchical view are hyper-linked to their respective dictionary entries in the second pane. Graphical depictions of CWE content, as well as the contributing sources, are also available on the site. Finally, the xml and xsd for the CWE List are provided for those who wish to do their own analysis/review with other tools. Dot notation representations of this material will be added in the future.

4. EXPANDING CWE

With the current draft of CWE List as a baseline/reference point, we are now gathering in the specific details and descriptions of 13 organizations that have agreed to contribute their intellectual property to the CWE initative. Under Non-Disclosure Agreements with MITRE, which allow the merged collection of their individual contributions to be publicly shared in the CWE List, Cenzec, Core Security, Coverity, Fortify, Interoperability Clearinghouse, Klocwork, Ounce Labs, Parasoft, proServices Corporation, Secure Software, SPI Dynamics, Veracode, and Watchfire are all contributing.

In addition to these sources, we will also leverage the work, ideas, and contributions of researchers at Carnegie Mellon's CERT/CC, IBM, KDM Analytics, Kestrel Technology, MIT Lincoln Labs, North Carolina State University, Oracle, the Open Web Application Security Project (OWASP), Security Institute, UNISYS, the Web Application Security Consortium (WASC), Whitehat Security, and any other interested parties that wish to contribute.

We expect the merging and combining of the contributed materials will take most of the summer and result in an updated CWE List that will be ready for community comments and refinement as we move forward. A major part of this will be refining and defining the required attributes of CWE elements into a more formal schema defining the metadata structure necessary to support the various uses of CWE List. This schema will also be driven by our need to align with and support the SAMATE and OMG SwA SIG efforts that are developing software metrics, software security tool metrics, the software security tool survey, the methodology for validating software security tool claims, and the reference datasets.

5. CURRENT THOUGHTS ON IMPACT AND TRANSITION OPPORTUNITIES

As stated in the concept paper that laid out the case for developing the CWE List [8], the completion of this effort will yield consequences of three types: direct impact and value, alignment with and support of other existing efforts, and enabling of new follow-on efforts to provide value that is not currently being pursued.

Following is a list of the direct impacts this effort will yield. Each impact could be the topic of much deeper and ongoing discussion.

1. Provide a common language of discourse for discussing, finding and dealing with the causes of software security vulnerabilities as they are manifested in code, design, or architecture.

2. Allow software security tool vendors and service providers to make clear and consistent claims of the security weaknesses that they cover to their potential user communities in terms of the CWEs that they look for in a particular code language. Additionally, a new "CWE Compatibility" will be developed to allow security tool and service providers to publicly declare their capability's coverage of CWEs.

3. Allow purchasers to compare, evaluate and select software security tools and services that are most appropriate to their needs including having some level of assurance of the level of CWEs that a given tool would find. Software purchasers would be able to compare coverage of tool and service offerings against the list of CWEs and the programming languages that are used in the software they are acquiring.

4. Enable the verification of coverage claims made by software security tool vendors and service providers (this is supported through CWE metadata and alignment with the SAMATE reference dataset).

5. Enable government and industry to leverage this standardization in the contractual terms and conditions.

Following is a list of alignment opportunities with existing efforts that are provided by the results of this effort. Again, each of these items could be the topic of much deeper ongoing discussion.

1. Mapping of CWEs to CVEs. This mapping will help bridge the gap between the potential sources of vulnerabilities and examples of their observed instances providing concrete information for better understanding the CWEs and providing some validation of the CWEs themselves.

2. Bidirectional alignment between the common weaknesses enumeration and the SAMATE metrics effort.

3. Any tool/service capability measurement framework that uses the tests provided by the SAMATE Reference Dataset would be able to leverage this common weakness dictionary as the core layer of the framework. This framework effort is not an explicitly called out item in the SAMATE charter but is implied as necessary to meet the project's other objectives.

4. The SAMATE software security tool and services survey effort would be able to leverage this common weaknesses dictionary as part of the capability framework to effectively and unambiguously describe various tools and services in a consistent apples-to-apples fashion.

5. There should be bidirectional alignment between this source of common weaknesses and the SAMATE reference dataset effort such that CWEs could reference supporting reference dataset entries as code examples of that particular CWE for explanatory purposes and reference dataset entries could reference the associated CWEs that they are intended to demonstrate for validation purposes. Further, by working with industry, an appropriate method could be developed for collecting, abstracting, and sharing code samples from the code of the products that the CVE names are assigned to with the goal of gathering these code samples from industry researchers and academia so that they could be shared as part of the reference dataset and aligned with the vulnerability taxonomy. These samples would then be available as tailoring and enhancement aides to the developers of software assessment security tools. We could actively engage closed source and open source development organizations that work with the CVE initiative to assign CVE names to vulnerabilities to identify an approach that would protect the source of the samples while still allowing us to share them with others. By using the CVE-based relationships with these organizations, we should be able to create a high-quality collection of samples while also improving the accuracy of the software product security assessment tools that are available to the software development groups to use in vetting their own product's code.

6. Any validation framework for tool/service vendor claims, whether used by the purchasers themselves or through a 3rd

party validation service, would rely heavily on this common weakness dictionary as its basis of analysis. To support this, we would work with researchers to define the mechanisms used to exploit the various CWEs for the purposes of helping to clarify the CWE groupings and as a possible verification method for validating the effectiveness of the tools that identify the presence of CWEs in code by exploring the use of several testing approaches on the executable version of the reviewed code. The effectiveness of these test approaches could be explored with the goal of identifying a method or methods that are effective and economical to apply to the validation process.

7. Bidirectional mapping between CWEs and Coding Rules, such as those deployed as part of the DHS NCSD "Build Security In" (BSI) website [9], used by tools and in manual code inspections to identify common weaknesses in software.

8. Leveraging of the OMG technologies to articulate formal, machine parsable definitions of the CWEs to support analysis of applications within the OMG standards-based tools and models.

Following is a list of new, unpursued follow-on opportunities for creating added value to the software security industry.

1. Expansion of the Coding Rules Catalog on the DHS BSI website to include full mapping against the CWEs for all relevant technical domains.

2. Identification and definition of specific domains (language, platform, functionality, etc.) and relevant protection profiles based on coverage of CWEs. These domains and profiles could provide a valuable tool to security testing strategy and planning efforts.

With this fairly quick research and refinement effort, this work should be able to help shape and mature this new code security assessment industry, and dramatically accelerate the use and utility of these capabilities for organizations and the software systems they acquire, develop, and use.

6. ACKNOWLEDGMENTS
The work contained in this paper was funded by DHS NCSD.

7. REFERENCES

[1] "The Software Assurance Metrics and Tool Evaluation (SAMATE) project," National Institute of Science and Technology (NIST), (http://samate.nist.gov).

[2] "The OMG Software Assurance (SwA) Special Interest Group," (http://swa.omg.org).

[3] "The Common Weaknesses Enumeration (CWE) Initiative," MITRE Corporation, (http://cve.mitre.org/cwe/).

[4] "The Preliminary List Of Vulnerability Examples for Researchers (PLOVER)," MITRE Corporation, (http://cve.mitre.org/docs/plover/).

[5] "The Common Vulnerabilities and Exposures (CVE) Initiative," MITRE Corporation, (http://cve.mitre.org).

[6] Viega, J., The CLASP Application Security Process, Secure Software, Inc., http://www.securesoftware.com, 2005.

[7] McGraw, G., Chess, B., Tsipenyuk, K., "Seven Pernicious Kingdoms: A Taxonomy of Software Security Errors". "NIST Workshop on Software Security Assurance Tools, Techniques, and Metrics," November, 2005 Long Beach, CA.

[8] Martin, R. A., Christey, S., Jarzombek, J., "The Case for Common Flaw Enumeration". "NIST Workshop on Software Security Assurance Tools, Techniques, and Metrics," November, 2005 Long Beach, CA.

[9] Department of Homeland Security National Cyber Security Division's "Build Security In" (BSI) web site, (http://buildsecurityin.us-cert.gov).

A Proposed Functional Specification for Source Code Analysis Tools

Michael Kass, Michael Koo, Paul E. Black, Vadim Okun
National Institute of Standards and Technology
100 Bureau Drive, Mail Stop 897
Gaithersburg MD, 20899

Abstract:

Software assurance tools are a fundamental resource for providing an assurance argument for today's software applications throughout the software development lifecycle (SDLC). Software requirements, design models, implementation code and executable code are analyzed by tools to determine if an application is truly secure. This document specifies the functional behavior of one class of software assurance tool: the source code analyzer. Because the majority of software weaknesses today are introduced at implementation, a specification that defines a "baseline" source code analysis tool capability can help software professionals select a tool that will meet their software assurance needs.

1. Introduction:

This section gives some technical background, defines terms we use in this specification, explains how concepts designated by those terms are related, and details some challenges in source code analysis for security assurance.

No amount of analysis and patching can imbue software with high levels of security or quality or correctness or other important properties. Such properties must be designed in and built in. Good choice of language, platform, and discipline are worth orders of magnitude more than reactive efforts. Nevertheless testing or examination of code has benefits in some situations.

Code must be analyzed to determine how different methods or processes affect the quality of the resultant code. If the origin of code has limited visibility, testing or static analysis are the only ways to gain higher assurance. Existing, legacy code must be examined to assess its quality and determine what, if any, remediation is needed.

Testing, or dynamic analysis, has the advantage of examining the behavior of software in operation. In contrast, only static analysis can be expected to find malicious trapdoors. Analysis of binary or executable code, including "bytecode," avoids assumptions about compilation or source code semantics. Only the binary may be available for libraries or purchased software. However, source code analysis can give developers feedback on better practices. Remediation is often done in source code. Analysis of higher level constructs, such as models, designs, use cases, or requirements documents, is possible, too. However, these higher level artifacts often lack rigor and rarely reflect all the critical detail in source code implementations. Thus static analysis of source code is a reasonable place to work for higher software assurance.

Often, different terms are used to refer to the same concept in software assurance and security literature. Different authors may use the same term to refer to different concepts. For clarity we give our definitions. To begin any event which is a violation of a particular system's explicit (or implicit) security policy is a *security failure*, or simply, failure. For example, if an unauthorized person gains "root" or "admin" privileges or if Social Security numbers can be read through the World Wide Web by unauthorized people, security has failed.

A *vulnerability* is a property of system security requirements, design, implementation, or operation that could be accidentally triggered or intentionally exploited and result in a security failure. (After [NIST SP 800-27]) In our model the source of any failure is a latent vulnerability. If there is a failure, there must have been a vulnerability. A vulnerability is the result of one or more *weaknesses* in requirements, design, implementation, or operation.

In the unauthorized privileges example above, the combination of the two weaknesses of allowing weak passwords and of not locking out an account after repeated password mismatches allow the vulnerability. This vulnerability can be exploited by a brute force attack to cause the failure of an unauthorized person gaining elevated privileges. An SQL injection vulnerability might be exploited several different ways to produce different failures, such as dropping a table or revealing all its contents. If spyware can steal a user's password, it is a vulnerability. But it may be hard to attribute the vulnerability to particular weaknesses in software that can be "fixed." Spyware typically exploits system weaknesses, which require changes at the system level.

Sometimes a weakness cannot result in a failure, in which case it is not exploitable and not a vulnerability. Such a weakness may be masked by another part of the software or it may only cause a failure in combination with another weakness. Thus we use the term "weakness" instead of "flaw" or "defect."

A source code analysis tool examines software and reports weaknesses or vulnerabilities it finds. They may be graded according to severity, potential for exploit, certainty that they are result in vulnerabilities, etc. Ultimately people must use the reports to decide

- which reported items are not true vulnerabilities,
- which items are acceptable risks and will not be mitigated, and
- which items to mitigate, and how to mitigate them.

The report may even lead the user to reject a piece of software altogether as insufficiently secure to use or as needing to be discarded and written from scratch.

For several reasons no tool can correctly determine in every conceivable case whether or not a piece of code has a vulnerability. First, a weakness may result in a vulnerability in one environment, but not in another. Second, Rice proved that no algorithm can correctly decide whether or not a piece of code has a property, such as a weakness, in every case. Third, practical analysis algorithms have limits because of performance and intellectual investment. Some vulnerabilities can only be identified if a tool performs inter-file, inter-procedural, or flow-sensitive analysis of the code. Deliberate obfuscation with complex code structures make the analysis even harder. Fourth, a tool may not have "rules" to find all known vulnerabilities. This is even harder since new exploits and vulnerabilities are being invented all the time.

Since no tool can be perfect, a tool may be biased on the side of caution and report questionable constructs. Some of those may turn out to be false alarms or *false positives*. To reduce time wasted on false alarms, a tool may be biased on the side of certainty and only report constructs which are (almost) certainly vulnerabilities. In this case it may miss some vulnerabilities. A missed vulnerability is called a *false negative*. Changing the threshold of certainty to report a construct as a vulnerability trades fewer false negatives for more false alarms and vice versa. The ideal would be a tool that reports every real vulnerability (no false negatives) with no false alarms. Even though this is theoretically impossible, utility requires some metric for the tradeoff between false alarms and false negatives.

2. Functional Requirements for Source Code Analysis Tools

In this section we first give a high-level description of the functional requirements for source code analysis tools, and then detail the mandatory and optional requirements.

High Level View

A baseline level of functionality is required in order for a source code analysis tool to be considered compliant with this specification. In its "simplest" sense, a source code analysis tool must be able to (at a minimum):

- Identify a select set of software security weaknesses in source code.
- Generate a text report of the security weaknesses that it finds, indicating the source file name and line number(s) where those weaknesses are located.

Requirements for Mandatory Features

In order to meet this baseline capability, all source code analysis tools must be able to accomplish the tasks described in the mandatory requirements listed below. If the tool under test supports the applicable feature, then optional requirements can be tested as well. If a specific tool does not provide the capabilities of a particular optional requirement, then the tool is not tested for that optional requirement. This means that a specific tool might provide none of the capabilities described under optional requirements. The following requirements are mandatory and shall be met by all source code analysis tools.

SCA-RM-1: The tool shall identify any code security weakness that is listed in appendix A.
SCA-RM-2: The tool shall generate a text report identifying all security weaknesses that it finds.
SCA-RM-3: The tool shall identify a weakness by its proper Common Weakness Enumeration [CWE] identifier.
SCA-RM-4: The tool shall specify the location of a weakness by providing the directory path, file name and line number.
SCA-RM-5: The tool shall be capable of detecting weaknesses within the coding constructs listed in appendix B.
SCA-RM-6: The tool shall generate an acceptably low "false-positive" ratio.

Requirements for Optional Features

The following requirements define optional tool features. If a tool provides the capability defined, the tool is tested as if the requirement were mandatory. If the tool does not provide the capability defined, the requirement does not apply.

SCA-RO-1: The tool shall produce an XML-formatted report.
SCA-RO-2: The tool shall have a "suppression system" that permits the user to identify and flag lines of code such that subsequent scans of the same (or modified) code will not generate the same report of a weakness.

Appendix A: Source Code Weaknesses

The source code weaknesses listed in this table represent a "base set" of code weaknesses that a source code analysis tool (or combination of source code analysis tools) should be able to identify if they support the analysis of the language in which the weakness exists. Criteria for selection of weaknesses include:

Found in real code today – The weaknesses listed below are found in real software applications.
Recognized by tools today - Tools today are able to identify these weaknesses in source code and identify their associated file names and line numbers.
Likelihood of exploit is medium to high – The weakness is fairly easy for a malicious user to recognize and to exploit.

Because the body of known software weaknesses is evolving (with new ones discovered every day), this list will grow. Additionally, as source code analysis tools mature in their capabilities and are able to identify more software weaknesses, those weaknesses will be added to this list. The names and descriptions in this list are found in [CWE].

Name	Description	Language(s)
Data Handling.Input Validation.Pathname Traversal and Equivalence Errors. Path Equivalence.		
Path Manipulation	Allowing user input to control paths used by the application may enable an attacker to access otherwise protected files.	C, C++, Java, other
Data Handling.Input Validation.Injection.		
Command Injection	Command injection problems are a subset of injection problem, in which the process is tricked into calling external processes of the attacker's choice through the injection of control-plane data into the data plane.	C, C++, Java, other
Cross Site Scripting.Basic XSS	'Basic' XSS involves a complete lack of cleansing of any special characters, including the most fundamental XSS elements such as "<", ">", and "&".	C,C++, Java, other
Resource Injection	Allowing user input to control resource identifiers might enable an attacker to access or modify otherwise protected system resources.	C, C++, Java, other
Data Handling.Input Validation.Injection.Command Injection.		
OS Command Injection	Command injection problems are a subset of injection problem, in which the process is tricked into calling external processes of the attacker's choice through the injection of control-plane data into the data plane. Also called "shell injection".	C, C++, Java, other
SQL Injection	SQL injection attacks are another instantiation of injection attack, in which SQL commands are injected into data-plane input in order to effect the execution of predefined SQL commands.	C, C++, Java, other
Data Handling.Range Errors.Buffer Errors.Unbounded Transfer ('classic overflow').		
Stack overflow	A stack overflow condition is a buffer overflow condition, where the buffer being overwritten is allocated on the stack (i.e., is a local variable or, rarely, a parameter to a function).	C, C++
Heap overflow	A heap overflow condition is a buffer overflow, where the buffer that can be overwritten is allocated in the heap portion of memory, generally meaning that the buffer	C, C++

	was allocated using a routine such as the POSIX malloc() call.	
Write-what-where condition	Any condition where the attacker has the ability to write an arbitrary value to an arbitrary location, often as the result of a buffer overflow.	C, C++
Format string vulnerability	Format string problems occur when a user has the ability to control or write completely the format string used to format data in the printf style family of C/C++ functions.	C, C++
Improper Null Termination	The product does not properly terminate a string or array with a null character or equivalent terminator. Null termination errors frequently occur in two different ways. An off-by-one error could cause a null to be written out of bounds, leading to an overflow. Or, a program could use a strncpy() function call incorrectly, which prevents a null terminator from being added at all. Other scenarios are possible.	C, C++

API Abuse.

Heap Inspection	Using realloc() to resize buffers that store sensitive information can leave the sensitive information exposed to attack because it is not removed from memory.	C, C++
Often Misused: String Management	Functions that manipulate strings encourage buffer overflows.	C, C++

Security Features.Password Management.

Hard-Coded Password	Storing a password in plaintext may result in a system compromise.	C/C++, Java

Time and State.Race Conditions.

Time-of-check Time-of-use race condition	Time-of-check, time-of-use race conditions occur when between the time in which a given resource (or its reference) is checked, and the time that resource is used, a change occurs in the resource to invalidate the results of the check.	C, C++, Java, other

Error Handling.

Unchecked Error Condition	Ignoring exceptions and other error conditions may allow an attacker to induce unexpected behavior unnoticed.	C, C++, Java, other

Code Quality.

Memory leak	Most memory leaks result in general software reliability problems, but if an attacker can intentionally trigger a memory leak, the attacker might be able to launch a denial of service attack (by crashing the program) or take advantage of other unexpected program behavior resulting from a low memory condition.	C, C++
Unrestricted Critical Resource Lock	A critical resource can be locked or controlled by an attacker, indefinitely, in a way that prevents access to that resource by others, e.g. by obtaining an exclusive lock or mutex, or modifying the permissions of a shared resource. Inconsistent locking discipline can lead to deadlock.	C, C++, Java, other
Double Free	Calling free() twice on the same value can lead to a buffer overflow.	C, C++
Use After Free	Use after free errors sometimes have no effect and other times cause a program to crash.	C, C++

Code Quality.Channel and Path Errors.Untrusted Search Path.

Uninitialized variable	Most uninitialized variable issues result in general software reliability problems, but if attackers can intentionally trigger the use of an uninitialized variable, they might be able to launch a denial of service attack by crashing the program.	C, C++
Illegal Pointer Value	This function can return a pointer to memory outside of the buffer to be searched. Subsequent operations on the pointer may have unintended consequences.	C, C++
Use of sizeof() on a pointer type	Running sizeof() on a malloced pointer type will always return the wordsize/8.	C, C++
Unintentional pointer scaling	In C and C++, one may often accidentally refer to the wrong memory due to the semantics of when math operations are implicitly scaled.	C, C++
Improper pointer subtraction	The subtraction of one pointer from another in order to determine size is dependant on the assumption that both pointers exist in the same memory chunk.	C, C++
Unsafe Reflection	By leveraging reflection capabilities, an attacker may be able to create unexpected control flow paths through the application.	Java

	potentially bypassing security checks.	
Null Dereference	Using the NULL value of a dereferenced pointer as though it were a valid memory address	C, C++

Encapsulation.

Private Array-Typed Field Returned From A Public Method	The contents of a private array may be altered unexpectedly through a reference returned from a public method.	Java, C++
Public Data Assigned to Private Array-Typed Field	Assigning public data to a private array is equivalent giving public access to the array.	Java, C++
Overflow of static internal buffer	A non-final static field can be viewed and edited in dangerous ways.	Java, C++
Leftover Debug Code	Debug code can create unintended entry points in an application.	C, C++, Java, other

Appendix B: Code Complexity Variations

In addition to having the capability to locate and identify source code weaknesses listed in appendix A, a source code analysis tool must be able to find those weaknesses within complex coding structures. A general list of these types of structures, adopted and modified from [MIT] is provided below. Some of the enumerated values are language specific (e.g. the use of pointers in C, C++), however, most are general types of constructs that exist across C/C++ and Java. Equivalent constructs in other languages will be added as tools for those languages are included in this specification.

Complexity	Description	Enumeration
address alias level	level of "indirection" of buffer alias using variable(s) containing the address	1,2
array address complexity	level of complexity of the address value of an array buffer	constant, variable, linear expression, nonlinear expression, function return value, array content value
array index complexity	level of complexity of the index value of an array buffer using variable assignment	constant, variable, linear expression, nonlinear expression, function return value, array content value
array length/limit complexity	level of complexity of array length or limit value	constant, variable, linear expression, nonlinear expression, function return value, array content value
container	containing data structure	array, struct, union, array of structs, array of unions, class

Memory leak	Most memory leaks result in general software reliability problems, but if an attacker can intentionally trigger a memory leak, the attacker might be able to launch a denial of service attack (by crashing the program) or take advantage of other unexpected program behavior resulting from a low memory condition .	C, C++
Unrestricted Critical Resource Lock	A critical resource can be locked or controlled by an attacker, indefinitely, in a way that prevents access to that resource by others, e.g. by obtaining an exclusive lock or mutex, or modifying the permissions of a shared resource. Inconsistent locking discipline can lead to deadlock.	C, C++, Java, other
Double Free	Calling free() twice on the same value can lead to a buffer overflow.	C, C++
Use After Free	Use after free errors sometimes have no effect and other times cause a program to crash.	C, C++

Code Quality.Channel and Path Errors.Untrusted Search Path.

Uninitialized variable	Most uninitialized variable issues result in general software reliability problems, but if attackers can intentionally trigger the use of an uninitialized variable, they might be able to launch a denial of service attack by crashing the program.	C, C++
Illegal Pointer Value	This function can return a pointer to memory outside of the buffer to be searched. Subsequent operations on the pointer may have unintended consequences.	C, C++
Use of sizeof() on a pointer type	Running sizeof() on a malloced pointer type will always return the wordsize/8.	C, C++
Unintentional pointer scaling	In C and C++, one may often accidentally refer to the wrong memory due to the semantics of when math operations are implicitly scaled.	C, C++
Improper pointer subtraction	The subtraction of one pointer from another in order to determine size is dependant on the assumption that both pointers exist in the same memory chunk.	C, C++
Unsafe Reflection	By leveraging reflection capabilities, an attacker may be able to create unexpected control flow paths through the application.	Java

Null Dereference	potentially bypassing security checks. Using the NULL value of a dereferenced pointer as though it were a valid memory address	C, C++
Encapsulation.		
Private Array-Typed Field Returned From A Public Method	The contents of a private array may be altered unexpectedly through a reference returned from a public method.	Java, C++
Public Data Assigned to Private Array-Typed Field	Assigning public data to a private array is equivalent giving public access to the array.	Java, C++
Overflow of static internal buffer	A non-final static field can be viewed and edited in dangerous ways.	Java, C++
Leftover Debug Code	Debug code can create unintended entry points in an application.	C, C++, Java, other

Appendix B: Code Complexity Variations

In addition to having the capability to locate and identify source code weaknesses listed in appendix A, a source code analysis tool must be able to find those weaknesses within complex coding structures. A general list of these types of structures, adopted and modified from [MIT] is provided below. Some of the enumerated values are language specific (e.g. the use of pointers in C, C++), however, most are general types of constructs that exist across C/C++ and Java. Equivalent constructs in other languages will be added as tools for those languages are included in this specification.

Complexity	Description	Enumeration
address alias level	level of "indirection" of buffer alias using variable(s) containing the address	1,2
array address complexity	level of complexity of the address value of an array buffer	constant, variable, linear expression, nonlinear expression, function return value, array content value
array index complexity	level of complexity of the index value of an array buffer using variable assignment	constant, variable, linear expression, nonlinear expression, function return value, array content value
array length/limit complexity	level of complexity of array length or limit value	constant, variable, linear expression, nonlinear expression, function return value, array content value
container	containing data structure	array, struct, union, array of structs, array of unions, class

local control flow	type of control flow around weakness	if,switch,cond,goto/label,setjmp,longjmp, function pointer, recursion
data type	type of data read or written	character,integer,floating point,wide character,pointer,unsigned character,unsigned integer
asynchronous	asynchronous coding construct	threads, forked process, signal handler
index alias level	level of buffer index alias indirection	1,2
loop structure	type of loop construct in which weakness is embedded	standard for,standard do while, standard while, non standard for, non standard do while, non standard while
loop complexity	component of loop that is complex	initialization, test, increment
memory access	type of memory access related to weakness	read, write
memory location	type of memory location related to weakness	heap, stack, data region, BSS, shared memory
pointer	pointer used for a buffer address	yes,no
scope	scope of control flow related to weakness	same, inter-procedural, global,inter-file/inter-procedural, inter-file/global
taint	type of tainting to input data	argc/argv, environment variables, file or stdin, socket, process environment

Appendix C: References

[CWE] Common Weakness Enumeration, The MITRE Corporation, web site
http://cve.mitre.org/cwe/index.html#tree

[MIT] Kendra Kratkiewicz and Richard Lippmann, A Taxonomy of Buffer Overflow for Evaluating Static and Dynamic Software Testing Tools. In Proceedings of Workshop on Software Security Assurance Tools, Techniques and Metrics. NIST SP500-256. Feb. 2006.

[SP800-27] Engineering Principles for Information Technology Security (A Baseline for Achieving Security), NIST SP 800-27, Revision A, June 2004. Available at http://csrc.nist.gov/publications/nistpubs/

www.ingramcontent.com/pod-product-compliance
Lightning Source LLC
Chambersburg PA
CBHW081849170526
45167CB00007B/2940